SELLING THE AMERICAN PEOPLE

Distribution Matters
Edited by Joshua Braun and Ramon Lobato

The Distribution Matters series publishes original scholarship on the social impact of media distribution networks. Drawing widely from the fields of communication history, cultural studies, and media industry studies, books in this series ask questions about how and why distribution matters to civic life and popular culture.

Emily West, *Buy Now: How Amazon Branded Convenience and Normalized Monopoly*

Robin Steedman, *Creative Hustling: Women Making and Distributing Films from Nairobi*

Anne Kaun and Fredrik Stiernstedt, *Prison Media: Incarceration and the Infrastructures of Work and Technology*

Lee McGuigan, *Selling the American People: Advertising, Optimization, and the Origins of Adtech*

SELLING THE AMERICAN PEOPLE

ADVERTISING, OPTIMIZATION, AND THE ORIGINS OF ADTECH

LEE McGUIGAN

The MIT Press
Cambridge, Massachusetts
London, England

The MIT Press would like to thank the anonymous peer reviewers who provided comments on drafts of this book. The generous work of academic experts is essential for establishing the authority and quality of our publications. We acknowledge with gratitude the contributions of these otherwise uncredited readers.

This book was set in Stone Serif and Stone Sans by Westchester Publishing Services. Printed and bound in the United States of America.

Library of Congress Cataloging-in-Publication Data

Names: McGuigan, Lee, 1986– author.
Title: Selling the American people : advertising, optimization, and the origins of adtech / Lee McGuigan.
Description: Cambridge, Massachusetts : The MIT Press, [2023] | Series: Distribution matters | Includes bibliographical references and index.
Identifiers: LCCN 2022037239 (print) | LCCN 2022037240 (ebook) | ISBN 9780262545440 (paperback) | ISBN 9780262374231 (epub) | ISBN 9780262374248 (pdf)
Subjects: LCSH: Internet marketing—United States. | Internet advertising—United States. | Advertising—United States. | Digital media—United States. | Social media—United States.
Classification: LCC HF5415.1265 .M367 2023 (print) | LCC HF5415.1265 (ebook) | DDC 658.8/72—dc23/eng/20230103
LC record available at https://lccn.loc.gov/2022037239
LC ebook record available at https://lccn.loc.gov/2022037240

10 9 8 7 6 5 4 3 2 1

CONTENTS

ACKNOWLEDGMENTS

I suspect it's always hard to write a book. Writing one during a pandemic seemed especially tough, even with all the privileges I enjoy. This book would not exist without the extraordinary efforts of many wonderful people and a few women in particular.

I was very fortunate to begin this project under the guidance of incredible mentors. Joseph Turow, Oscar Gandy, Carolyn Marvin, and Victor Pickard each imprinted something special on my work through their care and generosity. Nothing I could write here would be adequate thanks. I am so grateful that Ramon Lobato and Josh Braun saw something in my dissertation that made them believe it could become a book. They invested in this project and supported me throughout the process. Justin Kehoe and Gita Manaktala at the MIT Press deserve many thanks for helping this take shape. And the anonymous reviewers who read a draft of the manuscript gave some of the most thoughtful feedback I've ever received.

I would not have gone far without help from fabulous librarians: Sharon Black, Min Zhong, Stephanie Willen Brown, and the many unseen people who answered requests for books, articles, photocopies, and more. Thanks to the interviewees who talked with me. Special thanks to Paul Woidke, who patiently taught me all about the cable business. Thanks to organizations that let me attend events for free or discounted rates. Thanks to Brian Kenny and the Cable Center's Barco Library for letting me visit its archives. And thanks to the Social Sciences and Humanities Research Council of Canada for funding my research.

Many friends and colleagues shared thoughts on versions of this work. Their feedback has been invaluable, even if unsuccessful at overcoming my stubborn mistakes. Thanks to Bridget Barrett, Scott Brennan, Matt Crain, MC Forelle, Jake Goldenfein, Rachel Kuo, Robin Mansell, Vincent Manzerolle, Fenwick McKelvey, Marcel Rosa-Salas, Dan Schiller, Aaron Shapiro, Salomé Viljoen, and Sarah Myers West.

My colleagues in the UNC Hussman School of Journalism and Media provided an incredible atmosphere for the last and hardest stages of writing. Daniel Kreiss, Shannon McGregor, Susan King, Heidi Hennink-Kaminski, Francesca Dillman Carpentier, and John Sweeney gave encouragement and support as I got situated in my new home. Folks at the world-class Center for Information, Technology, and Public Life (CITAP) have somehow been as kind as they are brilliant. And Bridget Barrett and LaRisa Anderson contributed crucial research help.

Everywhere I've gone I've met amazing people. Some have endured my friendship with impressive resilience. Disgruntled readers should blame them for enabling me. At Penn's Annenberg School for Communication: Opeyemi Akanbi, Omar Al-Ghazzi, Rosie Clark-Parsons, Nora Draper, Jasmine Erdener, Nick Gilewicz, Kevin Gotkin, Emily Hund, Sanjay Jolly, Anne Kaun, Corrina Laughlin, Deb Lui, Elena Maris, Litty Paxton, Monroe Price, and Christin Scholz. In the Faculty of Information and Media Studies at the University of Western Ontario: Kyle Asquith, Robert Babe, Edward Comor, James Compton, Nick Dyer-Witheford, Alison Hearn, Keir Keightley, Daniel Robinson, Sandra Smeltzer, and the late Jonathan Burston (whose lessons in parsimony have saved readers from a truly unbearable experience). And in the Digital Life Initiative at Cornell Tech: Michael Byrne, Sally Chen, Maggie Jack, Ido Sivan-Sevilla, Jessie Taft, and Helen Nissenbaum, who has given me so many incredible opportunities. My experiences at DLI profoundly shaped the direction of my research.

I've benefitted from presenting parts of this work at meetings of the AI Now Institute, the Canadian Communication Association, CITAP, Cornell's AI Policy and Practice initiative, the International Association for Media and Communication Research, the International Communication Association, the Union for Democratic Communications, and a Canadian Media Policy reading group. Some material in this book was published previously in Lee McGuigan, "Automating the Audience Commodity: The Unacknowledged Ancestry of Programmatic Advertising," *New Media & Society* 21, no. 11–12

(2019): 2366–2385, https://doi.org/10.1177/1461444819846449, copyright © Lee McGuigan 2019; and Lee McGuigan, "Selling Jennifer Aniston's Sweater: The Persistence of Shoppability in Framing Television's Future," *Media Industries* 5, no. 1 (2018).

Most importantly, I would be nothing without the support of my family—especially my parents, Diane and John, my in-laws, Karen and Tom, and my wife, Emily, the roots beneath my feet. Along with my brother, grandparents, aunts, and uncles, these people have made possible everything I do. My friends Joel and Chris provided many welcome diversions. And my kids, Clayton and Birdie, have given me the best possible motivation to finish this project: the chance to spend more time with them.

I Dreams and Designs
to Optimize Advertising

INTRODUCTION: A WORLD MARKETERS CAN COUNT ON

The advertising industry chases its dreams. At a 1961 meeting of the American Marketing Association (AMA), the group's incoming president told executives from some of the country's leading corporations about an ideal method for deciding how much money to spend on advertising. "It calls for adding dollar after dollar to the appropriation until the point is reached where the last dollar produced no increase in profit."[1] The speaker was Albert W. Frey, a business school professor and one of a growing number of marketing experts who were dissatisfied with irrational and inefficient conventions in advertising management. Marketing had become an increasingly central and expensive factor in the US economy, with higher stakes and stricter scrutiny weighing on advertising strategies. The whole circulation of commodities seemed ripe for the sort of rationalization administered in manufacturing. As a 1962 article in *Harvard Business Review* explained, "In marketing, which represents over 50% of U.S. economic activity, or more than $250 billion annually, the opportunities for productivity gains are so large and numerous that we practically stumble over them. And the *crème-de-la-crème* lies in advertising."[2]

It was time to clear a path to the future. Frey warned marketers against "hiding behind" comfortable excuses like the well-worn adage, "Yes, I know that half of our advertising dollars are wasted, but I can't determine which half."[3] He urged management to face the challenge of carefully accounting for the financial return on those advertising investments. It was an imposing challenge, of course. To actually execute the optimal method Frey described would require "that the sales effectiveness of any given expenditure is known and that relevant costs are applied." That was not a practical reality for most companies. But it was something they could aspire to reach. And the possibility of maximizing profit on each dollar spent was a seductive reward for any company able to determine the exact effects of its advertising. "A few advertisers do claim to have this information to a remarkable degree," Frey reported, "but they are very much the exception."[4]

This book explains how companies dreamed of becoming the exception and what the advertising industry built as it tried to make the exception into the rule. By the late 1950s, some of the most prominent organizations in the US marketing system were busy trying to take advantage of computerization and the management sciences nurtured during and after World War II. They hoped dramatic improvements in their ability to produce and process information would allow them to subject advertising operations, and even consumer behavior, to more intensive accounting, calculation, and control. With algorithms, automation, and an army of applied researchers and mathematics experts all enlisted in the capitalist sales effort, it seemed possible for decision making in advertising to achieve machinelike regularity—efficient, precise, predictive, and quantifiably rational. Parts of the advertising business began to remodel themselves in the image of disciplines such as operations research, using mathematical models "to more or less automatically project the past and present into the future," as Frey put it at the AMA meeting.[5] Operations research was developed most famously as a science of optimization in warfare; its paradigmatic application is figuring out how to use an available stock of bombs to destroy the most submarines.[6] Advertising saw its own reflection in that basic problem and went to work sizing up the commercial world for target practice.

The data-driven business cultures that animate digital advertising were incubated in the aftermath. Marketers learned to dream of optimization and to speak in the idiom of management science. A general commitment to these ambitions and techniques has been an enduring force throughout technological shifts within US capitalism, influencing design values and power dynamics in the many institutions and infrastructures that integrate advertising with various aspects of social life. Increasingly, that includes not just news, entertainment, and shopping but also education, employment, insurance, credit scoring—almost any context in which an assessment of someone's value enters into decisions about how to classify and treat them.[7] Many of the developments in media and technology that fire public debates today have been fueled by marketers' desire to become more sophisticated accountants and engineers of everyday life. Pockets of the advertising business have exploded in recent years as people have come to claim, and to believe, that their dreams have come true, that the exception has become the rule, that new media have delivered a world marketers can count on.

Welcome to the frontier horizons of advertising technology. Advertisers and the many companies they work with imagine that new digital systems arm them with the data and analytical resources they need to realize a stubborn ambition: reaching a desirable consumer with the perfect marketing intervention at just the right moment to achieve their objective. More importantly, they believe these resources will allow them to predict and determine the financial return attributable to each strategic decision. They want to minimize any doubt about whether their ads succeed in generating sales or influence and to discern ever-narrower differences in profit and risk when evaluating the many options for allocating their budgets and orchestrating persuasive campaigns. Likewise, the companies that sell advertising inventory—search engines, social media, streaming services, mobile application developers, cable companies, and website publishers—all hope to maximize yield on what they sell. They try to do that by identifying potentially valuable consumers wherever they can be observed, sorting them into categories that capture something about their expected behaviors, and surrounding them with promotional messages and shopping opportunities adapted to those expectations. Brokering transactions between these parties is a dense sector of intermediaries providing personal data, auction platforms, trading-desk software, systems integrations, and various surveillance and analytics tools, often branded as artificial intelligence, that promise to help firms make better decisions and improve the efficiency of their marketing. These intermediaries provide logistical, transactional, and record-keeping services that try to turn any user action, any data flow, any instance of exchange into a billable event and a claim about some documented past or probable future. At its core, digital advertising is an automated technoscience, capitalizing evidence of attention and intention, correlation and difference, desire and disgust, value and cost. Welcome to the mercenary wilderness of "adtech."

Advertising has been a powerful wedge for digital platform businesses to accumulate astronomical fortunes by exploiting data about consumers, audiences, and markets.[8] In some ways, this marks a seismic shift. Economists used to concede that there is no accounting for taste. Today, companies desperately try to account for almost any detail of consumers' lives that might reveal something about the economic value of a person and the bespoke populations they might be assigned to, at any moment, within a database or an identity graph. Tastes, habits, moods, and movements are just part of what marketers and adtech systems take into account as they calculate predictions and make rapid judgments about how to classify people, how to nudge them toward desired behaviors, and how to price each opportunity to nudge.[9] Markets for personal data and data-derived services are sprawling and frenetic. In addition to data brokers such as Acxiom and Experian, these markets include broadband and cellular providers like Comcast and AT&T, retailers like Walmart and Kroger, payment processors like Master-Card, electronics manufacturers like Samsung, just about any application on a smartphone, and thousands of unseen or lesser-known businesses.[10]

Google, Meta (which owns Facebook and Instagram), and Amazon lead the pack of digital advertising companies.[11] They produce endless facts or inferences about people's activities, emotions, associations, and inclinations and process them along statistical assembly lines. Leveraging investments in computing power, user acquisition, elite expertise, corporate takeovers of competitors, and political and legal protection, these companies dissect everyday life so that they can claim to "know," with even a tiny fraction of additional accuracy, the value of each consumer target and every marketing maneuver. Observing what people do and where they go with unmatched detail, the largest adtech platforms document and monetize minuscule moments of user attention or behavior. They also profile publishers and advertisers to extract more value from the billions of transactions they orchestrate every day.[12] The "heterogeneous engineers" of these complex arrangements are working to align digital media with particular visions of a data-driven society,[13] fueling ongoing efforts to disaggregate and re-sort populations in ways that seem useful for advertisers and profitable for the firms that package and distribute ads. These are technological dreams of optimization, and they are pervasive among the companies trying every day to sell you products or to sell you to marketers.[14]

Some analysts look at these developments and see more than just a shift. They see a decisive break from the past. Shoshana Zuboff blames tech giants like Google and Facebook for ushering in an unprecedented age of "surveillance capitalism."[15] She and others argue that a new logic of surveillance and prediction has perverted capitalism and disrupted the business of advertising. These analysts presume that advertising today looks nothing like the pre-internet world memorialized on the popular TV show *Mad Men*. As one writer put it in 2018, "Advertising, once a creative industry, is now a data-driven business reliant on algorithms."[16] Prominent commentators and executives, including the former CEO of Cambridge Analytica, like to say that advertising's "mad men" have been dethroned by tech-savvy "math men." Don Draper has drowned in a data deluge.[17]

This narrative of digital disruption is a dominant frame in public conversations about adtech, data, and surveillance capitalism. Without dismissing it entirely, I believe this diagnosis needs some radical revisions. Despite recent frenzies of excitement and anxiety, capitalist dreams of optimization were not conjured overnight. Across a much longer period than we typically appreciate, related efforts to predict and influence consumers' habits, and to package and sell audience attention, have channeled and amplified currents in surveillance, information technology, and behavioral and management sciences. Advertising's logistical and calculative operations, deeply tied to computerization in US business, have helped legitimate a data- and prediction-obsessed attitude in the capitalist sphere of circulation and in media industries. As platform companies have insinuated surveillance and profiling into more means of sociality, advertising's role in the political economy of communication has, in important ways, been extended, clawing its way toward some fundamental and long-held ideals.[18]

The commercial media that have organized the flow of everyday information and cultural imagery in the United States for more than a century were shaped around advertisers' desire to produce consumers and consumption—that is, to assemble affluent and upwardly mobile people, including wage earners with new needs and disposable income, and conduct them through flexible sets of habits, lifestyles, and ideologies that reproduce capitalist institutions and the power of dominant social groups.[19] Other values compete against this priority, of course, and its success is far from uniform; but historical and contemporary research shows that the production of consumers, both as actors and as information, has been an organizing principle across print, broadcasting, and electronic media.[20]

Building on work by Dallas Smythe, Oscar Gandy Jr., Joseph Turow, and many others, I argue that the business of producing consumers has contributed decisively to ongoing schemes for monitoring and managing individuals, populations, marketplaces, and more. Many advertising and media professionals are compelled to engage in technoscientific performances of efficiency and advantage; to project authority within and across their organizations, they continually appropriate and publicize new means of generating and using information. The business traffics in a steady stream of data about consumers, their projected value to interested firms, and their anticipated and confirmed responses to marketing interventions.

Contrary to popular belief, this traffic is not exclusive to Silicon Valley or internet advertising, although its expansion and acceleration, its specific computing infrastructures and mathematical methods, and its deeper entanglements with financial speculation add up to something particular.[21] Still, a careful explanation of how accumulation strategies centered around the datafication of existence came to define an era of history needs to consider the elements that made this alchemy possible. That means examining much earlier efforts to create markets that trade in traces of human behavior and attention, as well as some largely unknown efforts to provide scientific techniques for managing operations in those markets—and managing consumption itself. Throughout the history of modern advertising, and especially since the mid-twentieth century, corporations across industries have agitated for better data about consumers and media audiences and more systematic means of exerting and measuring influence.[22] Dreams of optimization in advertising reflect forceful demands from practically all of business enterprise, including the suppliers of computers and information systems. Perhaps most interestingly, these dreams were tangled up with formidable technoscience disciplines that muscled their way toward the center of US universities, corporations, and parts of government. Surveillance advertising took inspiration and many of its marching orders from management science.

Management science and operations research collectively describe an eclectic battery of techniques for making "rational" decisions under conditions of uncertainty.[23] These disciplines aspire to technocratic governance of complex systems, using formal and usually quantitative models to evaluate opportunities, to make selections from alternatives that are difficult to distinguish or compare, and to predict the payouts of different choices. Of particular interest is their fascination with mathematical methods of "optimum seeking."[24] A 1957 textbook describes how experts used these methods to coordinate marketing systems: "the aim of the operations researcher is to find an appropriate optimization formula. That is to say that he is offering analytical help to management in minimizing costs, maximizing results, or identifying the best possible pattern of activities in a complex operation."[25] Martin K. Starr, a professor at Columbia University and a consultant to one of the world's largest advertising agencies in the 1950s and 1960s, says simply that optimization is "the fundamental objective of management science."[26]

Optimization is also the dazzling and fundamental objective of digital advertising. Its appeal is irresistible in the abstract. Detached from the social and political matter of specifying what values should be maximized (and for whom), optimization just means *more perfect*. I argue that this is exactly what adtech represents. It stands for automated management science—a sleek technological promise of more efficient and rational governance of the internet's vast marketing machines. But the details of that promise are contentious, irreducible to the bloodless elegance of technical perfection. Optimization represents a highly adaptable set of discourses that various companies and experts have harnessed in pursuit of their own status, power, and profit. These are fluid rivalries within contradictory but more or less collectively defended industrial formations. Convincing displays of optimum seeking have helped companies capture market share from their rivals while also reinforcing cultural and political-economic commitments to the idea that media systems should be organized largely—perhaps primarily—to produce consumers and to make moments of attention into investment opportunities.

Optimization in advertising has politics and a history, after all.[27] That history paved the way for today's distinctively algorithmic mode of "knowing," enacting, and packaging consumers and audiences. Beginning in the 1950s, computerization and management science—both thoroughly social as well as technoscientific developments—remediated the culture and material conditions of calculation and market exchange in advertising. They made optimizing return on investment into the holy grail of progress, a north star for sporadic efforts to define and exploit technological affordances. The data-driven, surveillance-based advertising that powers digital environments today did not pervert advertising's traditional commercial logic; rather, it emerged from *within* that logic, as the internal movements of long-standing business processes became entangled with new information technologies and all the promises they represented. What we call adtech is an intimidating but often clumsy assemblage of computers, databases, code, metrics, models and algorithms, consumers and users, high- and low-wage workers, scoring and classification systems, technical standards, market mechanisms, professional loyalties, and apparently authoritative claims to expertise and truth.

These are management science machines designed to optimize the private but sometimes conflicting goals of marketers, marketing service providers, and media companies. Their history shows how commercial relations that were settled around compromises (e.g., paying to expose audiences to advertising messages, measured by ratings) both shaped and adapted to new forces that seemed capable of resetting those compromises closer to ultimate objectives (e.g., paying to produce consumption, measured by sales). It's a story of "particular human efforts to negotiate difficulties and seize opportunities," to borrow a phrase from Richard Ohmann.[28] And, like so much of capitalist innovation, the glitches and good-enough fixes that adtech promoters assure us are temporary—certain to disappear as technologies realize their full potential—turn out to be enduring features of a future-in-the-making that never arrives at its promised destination.

"SELLING THE AMERICAN PEOPLE"

The marriage of media and marketing systems in the United States was arranged long ago. This courtship heated up in the last quarter of the nineteenth century as manufacturers of branded goods, modern advertising agencies, department stores in metropolitan centers like Philadelphia, New York, and Chicago, and mail-order merchants serving outlying areas all got involved with the publishers of mass-distributed periodicals and daily newspapers. This dance continued throughout the struggles to define and control radio broadcasting, with big technology companies cutting in and a government chaperone calling tunes. It took some seduction and counseling to fasten advertising to radio, but the resulting knot was tied tight. Expectations that mass media exist as extensions of corporate governance and the capitalist sales effort presided from the outset over the "age of television."[29]

In a 1957 address to the San Francisco Ad Club, the director of merchandising and promotion for a TV station in Los Angeles evangelized the young medium in suggestive terms: "My contention is that any activity which occupies the American people six and seven hours a day cannot be by-passed by advertisers interested in *selling the American people*."[30] He joined a chorus of influential contemporaries proclaiming television's importance to a consumer economy. Just three years earlier, as television was being welcomed into most US homes, revered business theorist Peter Drucker suggested that new technologies had transformed the practice of corporate management.

Rather than supplying markets with goods, he argued, management "must create customers and markets by conscious and systematic work." Drucker regarded television advertising as a form of "automation" for producing consumers and their buying habits, and he insisted that technological advancements in marketing and distribution would be as significant to the industrial system as the mechanization of manufacturing.[31]

This book is about "selling the American people." When the station executive spoke that phrase, he was extolling the power of television to influence consumer choices—selling people on advertisers' products. But his particular syntax also evokes a set of processes through which media audiences and consumer populations are produced, packaged, and sold to advertisers—what we call audience manufacture or commodification. Turning the statement's flexibility to our advantage, the book examines both sides of this coin: the "attention merchants" selling audiences to advertisers, and the "choice architects" trying to shape consumer behavior.[32]

While it is obvious that these operations have exploited and reinforced the prominence of commercial media as economic and cultural institutions, we have more to learn about how the production of consumers, markets, and media audiences has depended on and amplified the importance of information, computing, and management sciences throughout broader political economies. The advertising industry is an organized field of knowledge production and data flows. Marketing and audience manufacture have hosted sustained, albeit opportunistic, encounters among forms of science and technology concerned with observing, recording, classifying, and evaluating the personal details and behavioral patterns deemed useful for defining countless consumer publics.[33] Advertising constructs claims about race, class, gender, age, and value not only by creatively encoding images about brands and the people who use them but also through its internal operations and its points of intersection with other institutions of social ordering. The process of distributing advertisements and planning and evaluating their effects is inseparable from the industrial and epistemological production of consumers.[34] Companies involved in selling the American people paved a crossroads for ongoing ventures to catalog social relations and to derive insights for tactical advantage. Almost ninety years ago, John B. Watson, a pioneer of behavioral psychology and a longtime employee at the world's leading advertising agency, offered a marketing principle that hinged on the power of strategic intelligence: "It is getting yourself into a position where you can predict the other fellow's behavior that puts you in command in a selling situation."[35] At least since then, and with increasing urgency, advertisers, their agencies, and many species of managers, analysts, and data brokers have scrambled to occupy that position, leaning on a scaffold of information technologies.[36]

Selling the American People helps us make sense of today's attention merchants and choice architects by getting to the bottom of a few key questions: How did technical experts working at the intersection of data processing and management sciences come to command the center of gravity in advertising and media industries? How did their ambition to remake marketing through mathematical optimization shape and reflect developments in digital technology? In short, where did adtech come from, and how did data-driven marketing become a discernible configuration mediating daily encounters with people, products, and public spheres? Political and business elites have located the answers to those questions in a technological revolution that began in the 1990s with the market-led popularization of the internet and the eventual rise of mobile and social media. Critical researchers have given us a much clearer picture than that. Defense spending, economic crises, and cultural trends all brought information and communication toward the forefront of capitalist societies in the late twentieth century.[37] Policy choices created a governance structure for the internet that centered commodification and private profit.[38] Advertising's appropriation of new data extraction capabilities solidified an infrastructure for identifying and profiling consumers online and across devices.[39] Cybersecurity and government intelligence put the legitimizing force of statecraft behind commercial surveillance.[40] And finance capital helped funnel money, hype, and professional energy into adtech, digital marketing, and data analytics.[41]

A complex web of relationships institutionalized surveillance advertising as the beating heart of the internet across the last three decades. This book provides a detailed historical prelude to these formations. It looks back to show how particular calculative cultures and futuristic visions allowed optimization to take root as a powerful ideology and performative repertoire in advertising and marketing. The business of selling the American people provided a theater for acting out dreams of rational management and a workshop for preparing the technical and organizational capacities that let surveillance advertising flourish. I refer to this trajectory of continuity and change as advertising's calculative evolution.

ADVERTISING'S CALCULATIVE EVOLUTION

During the 1950s and 1960s, at the same time advertising underwent a well-known "creative revolution," the industry was also rebuilding its knowledge and decision-making infrastructures around electronic data processing and algorithmic and actuarial techniques. Advertisers, agencies, and media companies accommodated their activities to more statistical ways of thinking about consumers and audiences and more computational forms of reasoning and justification. By the 1960s, these and other actors recognized possibilities to reorganize mass media around efficient, automated systems for managing consumers and exchanging audience commodities—and they were already enlisting digital technologies and mathematical formalisms to press on the limits of established business processes. Power roles inside and across organizations were gradually renegotiated to favor functions suitable for or supportive of numerical modeling and optimization. Consumer research, media planning and buying, direct marketing, and data and computing services were all elevated in stature as advertisers and their agencies tried to discriminate risk and profit potential more precisely—drawing inspiration, legitimacy, and personnel from war-tested sciences of decision making and control.

Explicitly or not, these calculative corners of the advertising business promoted a technoscientific and financialized definition of progress as they worked to build markets where more dynamic units of attention or behavior could be packaged into finely graded investment opportunities. Throughout the second half of the twentieth century, the commitment to more detailed accounting of consumers' past and probable behaviors moved toward the very core of the sales effort and the digital media landscape. The escalation of these strategies and styles of industrial truth-telling made way for both public life and the intimacies of personal experience to be penetrated by computational approaches to knowing, designing, and managing social worlds.

This calculative evolution shows how advertising and marketing tried to leverage computer-centric management sciences and how an ideology of optimization inflected their orientation to information and technology. It is the history of a future imagined by marketers as they constructed the affordances of impressive new tools and techniques. Responding both to the difficulties of coordinating increasingly data-intensive business operations and to perceived opportunities to advance toward the dream of determining advertising's impact on sales, participants in advertising animated visions of progress through a set of affordances, or potential abilities, that became central to the commercialization of interactive and internet-enabled media. Efforts to reengineer advertising around automation, personalized targeting, electronic shopping, and expanded capacities for documenting, analyzing, and assigning value to consumer behaviors all contributed to a certain structure of feeling at the horizon of digital capitalism. These uneven and often flawed attempts to optimize the business of selling the American people helped create the conditions of possibility for today's behavioral advertising.

Advertising and marketing are among the many institutions in US capitalism that have gravitated toward actuarial and accounting logics and embodied those values in technologies of automation, surveillance, and quantification. This is the historical context in which adtech should be analyzed. The term "adtech" is usually reserved for internet-based advertising companies, particularly those involved in programmatic advertising, where individual opportunities to serve advertisements are sold in automated auctions. Google, Meta, and Amazon are the giants of adtech, leading a crowded field of variably familiar (and reputable) companies such as Microsoft, Adobe, Oracle, Roku, LiveRamp, The Trade Desk, FreeWheel (Comcast), Criteo, and MediaMath. Despite its typically narrow connotations, I suggest that adtech is essentially about generating, processing, and coordinating flows of information and commerce. The term can refer to the entire range of knowledge infrastructures, logistical utilities, and technoscientific practices involved in selling the American people.

By analyzing adtech within a longer history of information technology and management science in advertising, we get a different picture from that offered by the conventional wisdom about the origins of the systems we know today and the dynamics and actors involved in building them. This is not to deny that advertising has changed; but it has changed into something certain people in the business started to imagine, quite vividly, in the 1950s and 1960s. For all the differences between now and then, the performances and discourses that have legitimized adtech in recent decades profoundly resemble the many little dramas staged more than half a century ago to showcase the power of management sciences and digital computers, with similar promises of rational efficiency draped in technoscientific expertise.

Overall, the advertising business—a cultural and political-economic institution connected to many of the organizations, devices, and environments that mediate everyday experiences and power relationships—gradually reconstructed itself around aspirations to optimize a consumer society. Its efforts to leverage data and technology have reinforced the authority of quantitative and computational decision making, shaped the architecture of media and market settings (including privatized public spaces), affected the circulation of ideas and stories, and legitimized a paradigm of behavioral nudging that tries to exploit inferences about people's habits, emotions, and cognition.[42] Google claims that it exists to organize the world's information; it is better thought of as a stack of advertising utilities that organizes the world's people according to their apparent economic worth for certain marketers. Safiya Umoja Noble sets our focus with piercing clarity when she writes, "We need a full-on reevaluation of the implications of our information resources being governed by corporate-controlled advertising companies."[43] Collective action against surveillance-based behavioral advertising has been sidetracked by debates that misconstrue the problem as one of balancing trade-offs between "privacy" (defined in individualistic or even cryptographic terms) and the efficiencies and pleasures said to emerge from data-driven personalization. As Noble's prompt suggests, the larger and more pressing issue is to grapple with specific relations of power involved in creating and managing publics, public spheres, and public life.

DIFFERENCE ENGINES: DISCRIMINATION IS THE POINT

Digital advertising's hazards are well documented. Adtech companies slurp and slosh personal data all over the internet; they siphon funds away from journalists and other creative workers; and they monetize and profit from hateful and dishonest communications, either willfully or because they are incapable of handling the magnitude of their business.[44] Adtech's externalities become literally toxic when we consider the ecological cost of powering machines that spend day and night crunching numbers, making guesses, and placing bets, trying to accelerate consumerism. Moreover, since corporations can deduct advertising expenses from their tax bills,[45] and since advertising transactions are structured in ways that let intermediaries collect fees from all sorts of services and events, adtech facilitates an upward transfer of wealth toward certain classes of media and marketing professionals, including fraudsters.

Digital advertising companies also have a shameful record of discrimination.[46] Facebook has been beset by lawsuits and backlash for violating laws meant to protect people against prejudicial disadvantage in particular categories of business activity. Audit studies have repeatedly found that Facebook's ad delivery systems selectively bias specific populations, even if advertisers' targeting parameters do not stipulate those actions. One group of researchers calls this "discrimination through optimization."[47]

It is well known that algorithms and artificial intelligence systems reflect and reproduce social power and injustice as they "learn" from data that materialize existing relationships of inequality.[48] Clearly, we need public institutions, backed by the force of law, to intervene in the governance of the platforms and systems corporations use to manage affairs with customers and workers. In this book I add empirical force to one particular argument in these discussions: Discrimination is not a side effect of adtech. Discrimination is the point.

The purpose of data-driven marketing is to identify and isolate differences in the value assigned to profiled consumers or specific advertising events. Adtech embodies this purpose in machines that act rapidly on recognized distinctions in the risk or revenue potential of alternative courses of action—splitting hairs as finely as economy dictates. The goal is to find profit opportunities that are invisible or incalculable to humans or too fleeting for manual transactions to exploit. With big-data surveillance, marketers try to gain a better view of individuals as well as to recognize patterns and anomalies that could have predictive power for their valuations and probabilistic bets. In other words, the array of variables that can be used to judge someone as similar to or different from others in a customized population or audience is vast, constantly shifting, and often far removed from how a person or group might define their own identities, interests, inclinations, and solidarities.[49] The principle behind data-driven advertising is essentially to increase corporate capacities to assess a range of barely distinguishable options and select whichever choice is most likely to yield or save two pennies instead of one. Whether adtech companies like Facebook deliver on these promises is debatable, but businesses of all stripes have committed to this worldview. There is no question that *optimization through discrimination* is what adtech is selling and what advertisers are buying.

It is not necessarily bad to recognize differences, of course. Equity, for example, is a way of accounting for different needs and abilities. But it matters whose interests and priorities direct this social sorting. The discrimination of race, gender, class, age, ability, sexuality, and other propensities or associations is an active element of contemporary capitalism. Advertising participates in efforts to differentiate people, or even components of an individual's life, according to their predicted economic or political worth and to maintain these inequalities for surplus extraction.[50] Adtech doesn't just *discern* differences that exist within an external ground truth; it effectively *divines* differences from a reality of its own making. Many digital media environments engineer situations that produce distinctions among consumers or marketing moments. Adtech packages these as differentiated investments, converting evidence of variation into actionable claims of potential value, formalized for automated decision making. Digital advertising's machine-market configurations flatten the social texture of needs, abilities, and identities into commensurable units of probability and profitability.

Importantly, then, optimization through discrimination is not only a disposition in advertising but also a practical and historical achievement. That achievement depends on the building and maintenance of complementary infrastructures. It requires *technical* capacities to identify and isolate message recipients so that individuals or groups can be addressed exclusively. And it requires *administrative* capacities to make that technical addressability meaningful within business institutions—to buy and sell advertising opportunities in ways that capitalize on precise and dynamic variances in the apparent value of consumer targets or advertising events, while preserving enough standardization to prevent market actors from being overwhelmed by transaction costs and complexities. These ever-evolving capacities constitute the problem space for adtech.

Adtech is an adaptable discrimination engine—a capitalist technoscience producing and exploiting profitable correlations and deviations and operationalizing difference as revenue potential. It is an extension of what Gandy calls the "panoptic sort," a "complex discriminatory technology" that "considers *all* information about individual status and behavior to be potentially useful in the production of intelligence about a person's economic value."[51] Crucially, though, that does not mean marketers see people in whole. These are selective and distorted accountings, derived from indices of identity and sociality that are taken to be predictive of behaviors that marketers or platforms can monetize.[52] This sorting starts from often implicit but always political design choices and assumptions about what it means to be a valuable consumer or citizen.[53]

TAKING EVERYTHING INTO ACCOUNT

Discrimination implies an eye for detail. A central vector in the decades-long construction of adtech has been the expansion and reformatting of marketers' capacities to account for the elements of reality they care about and to admit those details into strategic calculations. Although we tend to think of advertising and media as creative industries, some of their most important activities mobilize scientific, administrative, and logistical assemblages. This is especially true with the distribution of advertisements—the processes involved in determining what ads are served, when, to whom, and at what prices. Advertising and media industries have been adapting their institutions and infrastructures so that a more comprehensive view of consumers' past and probable behaviors can be meaningfully incorporated into claims about someone's identity, value, and susceptibility to influence. Adtech's profiling, prediction, and decision-support systems recommend whether or how to engage with individuals or "types" of people based on countless data inputs, including the places they go, the websites they visit, the devices and applications they use, the things they buy, and even biometric cues about their personalities or emotions (although simple proxies for race or gender may overpower other signals). These data provide a basis for configuring media users, sometimes just momentarily, within computable consumer categories.[54] Automated agents acting on behalf of the companies that buy and sell access to consumers process these data and continually adjust their estimates of how much each advertising opportunity is worth. Surveillance, analytics, and algorithmic trading in advertising all help expand the "frame" of rationality—increasing the inventory of events, attributes, and relationships that can be taken into account when assessing choices, making decisions, and executing transactions.

The whole project of using management science to optimize advertising starts from these efforts to better account for consumers' behaviors and profitability. The goal, in a sense, is to enclose the sphere of consumption within the scope of corporate management.[55] That way, consumers—who exist outside of companies' direct control—can be incorporated into the quantitative models designed to represent and inform marketing processes. These models format consumers as probabilities and patterns that can be condensed into commensurable units of value, leveraging what Luke Stark calls "an investment in applying calculability to human subjectivity."[56] Digital advertising depends on accounting methods that make consumers calculable.

Computerization has shaped the backdrop for this strategic science. New information technologies seem capable of multiplying the phenomena available for measurement, increasing the precision of the resulting data, and amplifying the power to extract actionable insights. This is a threefold expansion of what counts: more sensors are installed in the world, more of reality is quantified, and more of that data matters in planning and evaluation. Not all accounts are treated equally, however. The preferences consumers "reveal" through observed behavior, or the propensities and identities that big-data analyses infer, are often considered more useful than anyone's own narrative accounts of who they are and what they want. For the most part, adtech's calculable consumers are produced by and for "machine intelligence." Individuals may not conform easily to this modeling and management, but failures have tended to motivate additional investments in invasive measurement and analysis.

Advertisers have committed to this elaborate project in part because they hope it will calm one of their deepest anxieties. The problem of *attribution*—isolating the relationship between advertising and consumer behavior—has haunted marketers at least since turn-of-the-twentieth-century magnates such as John Wanamaker worried that a sizable portion of their advertising had no measurable effect on profits. This uncertainty, and the dedication to taming it, inspires ongoing schemes for collecting data that could help marketers identify the people most likely to be valuable and then "close the loop" between advertisements and sales. The intention is to verify which marketing actions contributed to marketplace outcomes. Attribution implies both *taking* account (recording market-related events) and *giving* account (making claims about advertising effects and return on investment). The credibility of these accounts rests on the capacity to follow a consumer's path to purchase—or, at least, to know what ads a person saw and what things they bought. This means that these moments and events must be visible and available for analysis. Important parts of the digital media people engage with every day have been designed to generate massive evidence of consumer existence.

A historical look at attribution shows us that surveillance capitalism's data extraction imperative, as Zuboff calls it,[57] is not only about predicting behavior but also about enacting and confirming it. As Gandy points out, the instrumenting of marketplaces with intensive feedback mechanisms has served "to provide critical information about the extent to which advertising and promotional efforts have produced their desired effects on consumers."[58] Surveillance advertising reflects the dream of determining advertising's influence on sales—where "determine" means both to control and to ascertain.

Advertisers' attribution anxiety has sustained frantic aspirations of omniscience and stimulated demand for the information and services provided by ad agencies, management consultants, audience ratings firms, data brokers, today's platform giants, and the new retail media networks that leverage large merchants' customer relations and sales databases. One of the most alluring claims that Meta, Google, Amazon, and, to some extent, companies like Walmart make to advertisers is that they witness critical parts of the customer journey within their "walled gardens." They can give fuller, more convincing accounts—of behavior, value, effects, correlations, and probable futures—than other attention merchants and choice architects. Producing a constant stream of information about how people use media and behave in marketplaces, these platform companies live out a situation Marshall McLuhan saw taking shape. "The new human occupation of the electronic age has become surveillance," he said in a 1972 interview. "Whether you call it audience rating, consumer surveys and so on. . . . Espionage at the speed of light will become the biggest business in the world."[59]

DREAMS, DESIGNS, AND AFFORDANCES

A sociotechnical history of marketing and audience manufacture is a study in the imaginaries, institutions, and infrastructures of a data-intensive form of capitalism. Developments in advertising open a window into the expanded authority of computation and quantification in economic, political, and cultural life. Advertising was certainly not alone in this process, and it built on earlier and larger tendencies,[60] including surveillance and management techniques implemented to commodify enslaved people.[61] By the early twentieth century, some advertising professionals gazed longingly at the calculating ethos already developed in insurance, credit reporting, and other areas of American business that had been measuring, modeling, and risk-analyzing populations since the nineteenth century.[62] That ethos received an added technoscientific and geopolitical imprimatur after World War II. The hegemonic

thrust of capitalist democracy reframed labor-capital relations as a collaboration to expand participation in a consumer society where citizens were
defined by their purchasing power and marketplace choices. The production
and management of consumers linked America's largest corporations, the
advertising and public relations agencies they paid to engineer hospitable
political and economic cultures, emergent vendors of digital technologies
and information services, and a cadre of researchers employed by prestigious
universities and bankrolled from the deep pockets of the military-industrial
complex. Sciences of war were repackaged as authoritative but ostensibly
neutral sciences of optimization, and they were integrated into advertising
and marketing and the media industries they financed.

For well-placed actors, these changes in advertising's calculative capacities provided resources for telling stories about the future. Information
technology and management science became harbingers of progress toward
optimal economic activity. Caitlin Zaloom identifies a similar dynamic in
financial markets. She shows how the promise of new electronic equipment
created opportunities for trading spaces and procedures to be reconfigured
in ways that better approximated economic principles of efficiency and
rationality.[63] Likewise, from the late 1950s onward, advertisers, media buyers and sellers, and marketing researchers mutually accommodated computers and transactional routines to the goals of facilitating more rapid and
expansive data analysis and rationalizing decision making. Each exciting
advance created openings for advertisers to demand more and for agencies
and media to promise better.

This book showcases a succession of moments when advertisers or the companies serving them looked around and asked: How can we take advantage of
new technologies and sciences? How can we exploit an information revolution? Advertising professionals treated new technologies or scientific tools
as not-quite-blank screens for projecting capital's hopes and dreams, tinted
to accentuate the importance of their own expertise. Anything that seemed
to afford better knowledge and management of social relations could be animated as a mechanism for delivering efficiency, rationality, and optimization.

Across the second half the twentieth century, industrial discourses rallied
around a set of affordances that tell us something about how marketing—the
circulatory system of capital—viewed information, technology, and society.
Advertisers, agencies, and other intermediaries tried to define new technologies' potential by associating them with imagined abilities that stand out

now as the most salient features of digital adtech: programmability (auto-mation), addressability (discrimination and personalization), shoppability (interactive commerce), and accountability (measurement and analytics). The advertising industry exercised this repertoire to make sense of techno-logical possibilities across a range of settings, with mixed results. An archae-ology of these affordances—showing how they have been repeatedly and flexibly attached to new information technologies and techniques—offers insight into both continuities of logic and priority and new articulations of power. It provides a view of "disruption" that is not about the reversal or demise of an established order but about opportunities to seize a potential future where existing relations are deepened, extended, or accelerated.

CHAPTER DESCRIPTIONS

Chapter 1 describes the state of the art in digital advertising to illustrate some logistical and epistemological dimensions of adtech. Chapter 2 introduces the idea that advertising has undergone a calculative evolution and identifies an inflection point in that evolution in the 1950s and 1960s, associated with computerization and the influence of management sciences. The chapter then surveys earlier historical examples of calculative cultures in US adver-tising. Chapters 3 and 4 focus on the midcentury inflection point to show how operations researchers and management scientists sold advertisers on the promise of determining advertising effectiveness and helped agencies develop mathematical models to formalize decisions about where to adver-tise and how to spend clients' money. The four subsequent chapters exam-ine the set of affordances introduced above. These ideas and developments have been interwoven to some extent and are therefore threaded through-out the book. But each chapter is devoted to one of the potential abilities and the actors and activities orbiting it.

Chapter 5 historicizes programmability—efforts to automate media plan-ning and buying. From a midcentury management paradigm preoccupied with "systems" and "efficiency,"[64] many imagined a future where comput-ers and algorithms could be programmed to select advertising placements with speed and precision exceeding human capabilities. Automation made it seem possible to accelerate the flow of information and the pace of trans-actions. These ideas gave certain professionals and businesses the lever-age to assert power and expertise. In particular, this discourse channeled advertisers' existing demands for better data and more rational management and aligned those priorities with the technological and scientific authority of the media planning and research departments at large advertising agencies. Those departments accentuated the computational nature of their work, tethered their status to computers, and claimed to be best positioned to make the most of those expensive machines.

Chapter 6, on addressability, examines initiatives for producing an audience of one. Computerization and market segmentation combined to intensify efforts to zero in on the people expected to be receptive to advertisers' messages. By the 1970s and 1980s, cable TV operators were installing "addressable" systems that connected set-top devices in subscribers' homes to computer databases and control centers at cable distribution facilities. This allowed operators to identify and discriminate among individual households at a technical level. Addressability was intended to restrict unauthorized access to premium channels or exclusive content, but advertisers, cable advertising sales departments, and set-top box makers recognized the ability to isolate a specific device for targeted messaging and collect granular data about viewers. Addressability is fundamentally a technology of discrimination, specifying inclusion or exclusion. The chapter focuses on efforts by cable operators' sales representatives to market this capability to national advertisers—efforts which prefigured some foundational elements of internet advertising. Addressable cable advertising was a trial run in operationalizing the individual as a salable unit of audience.

Excitement about cable television and telecommunications also included a fixation on interactive shopping. Chapter 7 explores how shoppability has been used to frame the future of new media systems. Although laptops and mobile devices are the taken-for-granted gateways to e-commerce today, for thirty years, virtual shopping was imagined as the domain of interactive TV. This chapter documents efforts to engineer interactive and, specifically, transactive capabilities into the technologies and business models supporting entertainment and information services. Shoppability exemplifies what Vincent Mosco calls "pushbutton fantasies," discursive constructions that, "explicitly or not, seek to occupy the image space that people turn to when they think about what the new information technology means."[65] The prospect of "selling Jennifer Aniston's sweater," which became a recurring slogan in the advertising and television industries, encapsulates the pinnacle of this dream—that viewers could instantly purchase items appearing in programs and advertisements by clicking a button on their remote controls. The chapter shows that ideas about connectivity to electronic marketplaces surfaced repeatedly in discussions about the convergence of media technologies and industries. Although cycles of hype and frustration punctuate this history, "buy buttons" are now fixtures of online advertising, from YouTube to Instagram to TikTok. The shoppable feeds of popular influencers carry on the dream of selling Jennifer Aniston's sweater.

Chapter 8 examines accountability—meaning, literally, the ability to take into account. Because this theme has been a constant presence in talk about what information technologies make possible, the chapter spans the whole second half of the twentieth century. Calculating return on advertising investment is an overarching goal that motivates much of the strategy and action shaping adtech. Attribution—making claims about advertising's effects on sales—requires vast arrangements for collecting, processing, and analyzing information. These claims are rarely airtight, but the will to determine return on investment helps draw a dominant share of digital commerce toward Google, Meta, Amazon, and other companies that track people across the purchasing process. This is the production of consumers at its data-driven zenith. It is the sales effort reimagined as an automatic sales engine.

WHAT A NIGHTMARE

By now, readers may be feeling a desperate sense of foreboding. Both boosters and critics sometimes portray digital marketers as all-powerful puppet masters. But in many ways, the story that follows is one of ambition persisting through repeated disappointments. The list of debacles is long and growing. It is commonplace for web users to be stalked by ads promoting things they don't want or already bought. Programmatic advertising involves so much invisible human labor that critics call it "programmanual." Advertisers obsess over efficiency while they collectively lose tens of billions of dollars a year buying fraudulent inventory and paying mysterious fees to adtech vendors.[66] An early attempt at using interactive television to let viewers order pizza was so expensive to operate that one observer wondered whether it would have been cheaper to pay someone to bring pizzas to each subscriber's home and wait for them to get hungry.[67] Claims of optimization exist in a haze of marketing hype—from the service providers who want to lure clients and investors and from the clients who want to impress executives and shareholders. And, as Meta often demonstrates, companies can project a veneer of technocratic rationality while their leadership flails through a muck of politics and public relations.

The spectacular failure of schemes to optimize advertising has renewed existential reflection. Does advertising work at all? Does data-driven advertising work any better? For more than a century people have fought to answer, or avoid, these sorts of questions. Lately, some researchers and writers have claimed to expose the embarrassing secret that the effects of advertising are probably small, sometimes negative, and always difficult to prove. That's nothing new. The advertising industry has always assured worried publics— especially policymakers—that advertising informs but never manipulates people (although media companies and ad agencies tell much different stories to their customers). The novel claim from today's skeptics is that data-driven advertising, despite its high-octane artificial intelligence, is less powerful than people believe. It's mostly just snake oil, more artifice than intelligence.

This debate is interesting, sometimes scandalous, but mostly beside the point I want to make. Regardless of whether specific ads shape individuals' behaviors, the commitment to data-driven advertising shapes the social environments, the information environments, and even the built environments people encounter every day.[68] And it has done so for decades. The purpose of this book is not to answer the question does data-driven advertising work; it's to show that the dream of data-driven advertising is *doing* work. Efforts to predict and produce evidence of advertising effectiveness have motivated extensive investments in data extraction and analysis and helped cement surveillance and discrimination into the culture of modern business. The problem, in other words, is not just that Google and others claim to know what you will do next, but that countless companies are willing to pay them to find out. Adtech sells the ideal that Frey imagined back in 1961: with enough information, advertising experts can determine the return on every dollar they spend.

These commitments—to optimization, to identifying and exploiting finer profit opportunities, to making better predictions and decisions— have all helped build a world that marketers can count on: where media systems serve commercial priorities and quantify the elements of reality that all sorts of businesses and influencers want to manage. Chasing this dream has foreclosed other possible futures. Even if designs to optimize advertising feature plenty of pure fantasy, alternative ways of mediating social worlds have been overwhelmed by the habitual recruitment of digital technologies as means for accelerating the circulation of commodities and turning almost any observable signs of life into information products or assets. Despite all the skepticism about data-driven advertising's efficacy, that critical reflection has done little to unsettle the behavioral and management sciences that underwrite adtech from their place near the center of US research, policymaking, and business education. The economics profession hands out Nobel Prizes for the ideas adtech implements, and elite universities reproduce an influential class of optimum seekers, feeding graduates into tech companies that interlock with all sectors of the economy and the state and essentially sell management science as software platforms and information services. Critics who want to dismiss data-driven advertising as humbug need to contend with the larger, cross-institutional life of these ideas and practices. And while the growing currency of privacy as a regulatory issue and branding strategy may pour cold water on adtech's behavioral data gold rush, the zeitgeist is not finished with optimization, so struggles to control these circuits of information, culture, and capital will continue.

Surveillance capitalism is not inevitable. It has taken considerable effort to move society in this direction. But surveillance capitalism is not a disruptive reversal of the capitalist media and marketing systems of the twentieth century. It is an intensification of the orienting mission they have clung to for decades: to extend the optimizing power of management science to the commodification of everyday life—in other words, to produce consumers. The heinous social failures all around us go well beyond anything media reform can achieve on its own—failures of intersectional justice, environmental action, and democracy writ large. Adtech and social media are by no means singular causes of these cascading crises. Nevertheless, an organized response to current problems must include, among many political battles, a radical departure in the design, ownership, and control of media and information systems. Our means of communication, culture, and sociality need to be shaken from capitalist dreams of selling the American people and adapted to the messy but vital work of supporting public life. As we battle on, let's keep in mind Armand Mattelart's warning, still poignant thirty years later: "Marketing pursues its mad dream: to predict behaviour and maybe manage to control it. To penetrate the secret of the black box of the 'consumer.' For the future of the democracy of daily intercourse, one can only hope that the day on which they find the key . . . is far away."[69]

1   ADTECH FLOWS: CLAIMS, LOGISTICS, AND OPTIMIZATION IN DIGITAL ADVERTISING

Adtech is a major circulatory system of the digital economy. Marketers, media companies, and the intermediaries between them use an agglomeration of tools and infrastructures to route money and personal data around the world and around the clock. Information about internet users, predictions about their worth and future actions, and bids to serve advertisements ping-pong from computer to computer hundreds of billions of times a day. These flows of information and commerce animate huge swaths of the for-profit internet, helping some of the most highly capitalized companies in history collect almost all their revenue. In 2021 advertisers spent an estimated $211 billion on digital media in the United States and $491 billion worldwide. Analysts project those figures will increase by 50 percent and 60 percent, respectively, by 2025. Google, Meta, and Amazon dominate this trade, collecting almost two-thirds of US digital advertising revenue.[1] Thousands of lesser companies orbit these oligopolists, rigging tackle on advertising markets and fishing for scraps of money and data. Whether they are unknown, well known, or just notorious, adtech companies are fixtures of modern life. As Tim Hwang explains, "The need to create a liquid market in human attention influences the architecture of the social spaces of the web."[2]

Defining adtech is tricky. The term is used to characterize various means of automating and optimizing digital advertising processes. We could say that adtech is a collection of classification and decision systems running on proprietary data sets, algorithms, software platforms, and artificial intelligence products. The core operations in adtech cohere around two interrelated classes of activities: one is epistemological, involving predictions, conjectures, and knowledge claims (about value, identity, affinity, authenticity, and probability); the other is logistical, formatting and trafficking information (including HTTP requests, metadata about ads and ad inventory, server logs, sensor readings, identity tokens, user profiles, invoices, and receipts) and facilitating transactions at the inhuman speed and scale required for a business that trades a nearly unlimited volume of commodities that are each, in some sense, unique. Does the person using a website belong to a valuable consumer population? What is access to someone from that population worth at this moment? How likely is it that this user will click on a certain ad? Has the user seen the ad before? Will this ad placement be visible for more than one second? What happened after the ad was served? Answers to these questions are just some of the predictions and claims that are the stock-in-trade of adtech. A tangle of standards and logistics helps produce them in the blink of an eye. Adtech is an assemblage of machines, markets, mechanisms, mathematics, and human workers, all of them generating, processing, and coordinating flows of information and commerce.

It is, admittedly, imprecise to speak of adtech as something singular and coherent, even though that is the fashion. A discernible cultural identity has been erected around a subset of companies and specialists calling themselves adtech, yet I suggest the term should include any application of information technologies to mediate advertising transactions, to manage the distribution of ads, and to execute and evaluate decisions about what ads to serve, to whom, and at what prices. I resort to the singular noun as a matter of convenience, but I intend for this discussion to embrace a wide variety of businesses and operations.

Arguably, too, a critical definition of adtech should focus not only on existing markets, technologies, and practices but also on ever-evolving promises of progress. Adtech stands for a flexible universe of perceived solutions to advertising problems and ambitions. The specifics of these solutions reflect the range of strategies across firms and market sectors, and a procession of technoscientific trends cuts the emperor's robes to whatever style is in season, from big data to blockchain to the next attractive buzzword.

Whatever the details, the fundamental and overriding purpose is this: to recognize and act on differences in value. Or, more precisely, adtech enacts claims about value, creating and exploiting new investment opportunities. Advertisers, intermediaries, and attention merchants all try to use advantages based on information, speed, or analytical muscle to buy or sell units of audience attention and behavior more efficiently. Adtech comprises a loose cluster of difference engines trying to capture tiny but incremental fractions of surplus from billions upon billions of transactions. These engines are designed to identify or predict the profit potential of consumers and their moments of attention with competitive precision and thus increase the probability of achieving strategic objectives. In this sense, adtech is a kind of automated management science, an embodiment of decades-old dreams of remodeling advertising as a machinelike optimization system that orchestrates frictionless transactions and dynamic adaptations at the speed of light.

This chapter sketches the current adtech landscape. The sketch is necessarily incomplete. Adtech's economic and technical dimensions are vast, intricate, and constantly shifting. Even as I write, corporate and public policies related to data governance and privacy are forcing drastic changes in the business of collecting and monetizing personal information.[3] After decades of advocacy by critical researchers and civil society groups, popular and legal resistance to surveillance advertising is not only altering adtech processes but also foreshadowing diminished growth for the industry, which may depress stock prices and affect companies' strategic plans as investors channel their speculative capital elsewhere. Rather than trying to anticipate the future, I focus on current conditions, and I keep the discussion at a general level because existing techniques may abruptly fall into obsolescence or incompliance with the law. Still, I hope this chapter provides a descriptive map of adtech that helps readers situate themselves in the recent state of the art and navigate the historical contours traced throughout this book.[4]

BEHAVIORAL AND PROGRAMMATIC ADVERTISING

Adtech is typically associated with behavioral and programmatic advertising. Behavioral advertising means targeting media users with messages that have been selected or customized based on observations of or inferences about a person's actions and inclinations. It is sometimes called interest-based advertising, implying that ads reflect people's revealed or suspected interests, although it more accurately means that the targeted consumer is deemed interesting or relevant to the advertiser. The principal assumption here is that past behavior provides insight into an individual's habits and intentions and ultimately that person's value as a consumer, voter, or social actor. Data about lifestyle, location, buying patterns, psychographics, and more are assembled into profiles that then factor into automated or computer-supported decisions that determine the distribution and pricing of ads. These profiles include personal details that users disclose voluntarily (such as through social networking accounts), as well as passively collected records of behavior (such as clickstreams). Most commonly and straightforwardly, advertisers target internet users based on their internet usage: the websites they visit, the ads they click on, the items they put in their shopping carts at e-commerce sites, and so on. Behavioral advertising decisions can also draw on data generated by offline activities, such as car or home ownership, purchases made with credit cards or recorded through a store's loyalty rewards program, and much more.

Marketers also use statistical techniques to try to identify attributes a person has not disclosed or exhibited, to detect patterns of behavior in large data sets, and to predict future considerations—such as a customer's profitability to a firm, a person's probable responses to certain stimuli, or an individual's passage into a life stage associated with big purchases and new brand loyalties, like getting married or divorced. Frequently, marketers act on behavioral data in time-sensitive situations, such as a momentary assessment of emotion or a consumer's presence near a retail location.[5] The advertising industry has constructed elaborate technological scaffolds to help companies exploit micromomentary opportunities.

Most digital advertising depends on those scaffolds, using computers, algorithms, and auction mechanisms to automate and optimize decisions and transactions. This bundle of automation, optimization, and auctions is called programmatic advertising. "Programmatic advertising technology promises to make the ad buying system more efficient, and therefore cheaper," an editor at *Digiday* explains, "by removing humans from the process wherever possible. Humans get sick, need to sleep and come to work hungover. Machines do not."[6] Besides replacing "unreliable" human workers with stone-cold rational calculators, programmatic advertising also promises to unlock new efficiencies by dynamically assessing and extracting the precise value of each advertising opportunity. Instead of placing ads next to content that attracts certain types of people, advertisers buy individual *impressions*—opportunities to serve ads to users who are defined by some set of measured or estimated properties—wherever they come up for sale, irrespective of the surrounding content. As one executive put it, "Ads are being delivered individually, one at a time, to consumers based on data that identifies them as members of some ideal target audience an advertiser is seeking."[7] The capacity to single out consumer targets and to control the distribution of ads to exclude individuals or classes of people who do not fit a desired profile is called *addressability*.

Typically, what digital advertisers buy with their money is either an amount of exposure among a target audience or some measurable user action. In the first case, advertisers pay when an ad has been served and/or viewed by a user, and the price is set in terms of cost per mille (CPM), or the cost per one thousand impressions. Costs vary across advertising formats (e.g., display and video), but in general, programmatic advertising is quite cheap. With typical CPMs of just a few dollars, the winner of a programmatic display auction often pays only a few cents (or less) per impression. In contrast to these exposure-based transactions, some ads are sold on a cost-per-click (CPC) or cost-per-action (CPA) basis, where advertisers pay when a user clicks on an ad, downloads an application (app), signs up for an account, buys a product, visits a store, donates to a political campaign, and so on. In other words, advertisers are paying for documented events.

Different pricing models have implications for the distribution of ads. Google search ads, for example, are sold on a CPC basis. When an advertiser bids in a keyword auction for the opportunity to have its ad appear next to search results, Google's auction mechanism factors into that bid an estimate of the probability that a user will click on the ad. Because Google and other search engines (e.g., Microsoft's Bing) get paid when users click, these companies are keen to learn how to predict click-through rates. They employ and consult small armies of researchers who work at the intersection of computer, data, and management sciences to study and fine-tune the design of automated auctions. This is also true of Facebook. Advertisers pay the platform to orchestrate a wide variety of actions, from clicks to likes to customer acquisition. Facebook's primary service is using algorithmic decision systems to distribute ads into situations with an apparently high probability of achieving specified outcomes. Inflected by the causal language of attribution, these outcomes are often called conversions, and they can be selected as the objectives Facebook will try to optimize when making ad delivery choices for its customers. Facebook even allows advertisers to target users that the platform's machine learning technology predicts will generate the highest return on advertising dollars spent.[8]

Programmatic advertising can refer to a number of different configurations, depending on how deals are brokered, who has access to a supply of inventory, and which professionals and budgets are activated in the buying process. Auctions and algorithms are used to place addressable ads within social media feeds, on mobile apps, between songs on Spotify, and surrounding videos on YouTube, Hulu, and other streaming services. Even inventory on digital billboards and other "out-of-home" advertising screens is sold using similar methods.[9]

Perhaps the best-known type of programmatic advertising is the real-time bidding that occurs when internet users navigate to websites and thereby create opportunities to serve ads. In the few hundred milliseconds between the time a user clicks on a link or enters a URL into a browser and the site loads on the user's device, an auction takes place to determine which ad to insert. This process is facilitated by intermediary technology platforms that circulate information about the impression for sale, execute programmed or learned decision rules, and keep a portion of the money advertisers spend. Publishers—meaning any website, big or small, with advertising inventory to sell—have relationships with intermediaries that help them auction their impressions. For instance, most publishers employ an ad server, almost always run by Google, to manage the routing, insertion, and logging of advertisements. When a user visits a website, the user's browser contacts the site's server to request the content; it also contacts the site's ad server, telling it to populate any available ad slots. Publishers and ad servers often plug into supply-side platforms (SSPs) to access more sources of demand for their advertising inventory than would be feasible with just an internal sales staff. Initially, SSPs were decision support systems that predicted which advertising sales network was likely to return the best price for the publisher's inventory. These ad networks, operated by companies like Google and Facebook, function similarly to the way sales representatives in other media industries have functioned for generations—simplifying the logistics of making ad inventory from a large number of sellers accessible to many potential buyers.

SSPs now make publishers' inventory available to buyers by feeding into the next link in the chain—the advertising exchange. Ad exchanges conduct auctions by interconnecting with demand-side platforms (DSPs) that carry out bidding and campaign management functions for advertisers and ad agencies. Ad exchanges contact DSPs with bid requests that convey data about impressions and the users they represent (e.g., location, device type and operating system, and, potentially, a consumer profile attached to a unique user or device identifier). The DSP tries to recognize and buy the most valuable or efficient advertising opportunities, given its programmed objectives and parameters—such as whom to target, how much to spend, and what outcomes to optimize. As one industry expert put it, DSPs "receive on the order of 1 million opportunities to bid every second. And for each one of those opportunities they make a judgment: how much money is this opportunity worth to [the client's] brand?"[10] Multiple DSPs bid in programmatic auctions, and in many cases, each of them first conducts an internal auction to determine which client's bid to submit to the exchange. Ad exchanges quickly collect and adjudicate bids, select winners, and report the outcomes back to the ad servers, which then distribute the appropriate ads into the slots on the websites and trigger billing and record-keeping actions. Small advertisers also participate in programmatic auctions using self-service buying tools that function like DSPs.

Inputs for these valuations and decisions come from a variety of tracking and data integration tools. The cookies and pixels that websites and marketers use to monitor traffic, profile visitors, and measure advertising outcomes are the workhorses of surveillance advertising. A variety of additional tools operate at other layers of the infrastructure, such as application programming interfaces (APIs) and software development kits (SDKs), which enable the circulation of information about what people do on mobile apps and social media platforms. These tools facilitate integrations across data and analytics services, helping to stitch the seams of an ecosystem that aspires to envelop a multitude of sites and contexts.[11] Programmatic transactions thus depend on data brokers, data management platforms (DMPs), and identity resolution vendors to connect bits of information held by different parties. For example, many advertisers have first-party data stored in customer relationship management systems; DMPs match those data sets with the information publishers have about their audiences. Essentially, data and identity vendors help marketers determine and recognize the value of profiled consumers and thus calculate probabilities and profit potentials in particular auctions. The same is true on the other side of the auction, as suppliers of advertising inventory use DMPs to discern the likely market value of what they sell. By doing so, they may be able to set an auction reserve price—the "floor" below which they will not sell an impression—to maximize revenue.

The trading process is largely automated, but it still relies on human workers for direction and extensive help. For example, programmatic buying agencies employ people to specify and adjust bidding parameters, which then guide DSPs' algorithmic decisions. This work involves selecting options from drop-down menus or populating fields in a software interface. Despite the blanketing pretense of rationality, programmatic buyers sometimes exercise informal priorities, such as deviating from prescribed bidding guidelines in order to liquate any unspent portion of a client's budget before the scheduled close of a campaign. Staff at buying agencies also prepare, curate, or review documents that report (with varying degrees of precision, accuracy, and honesty) what clients got for their money. Elsewhere in the pipeline, people work at "cleaning" data sets, maintaining software products and system integrations, and managing business-to-business relationships. Dreams of payroll savings via automation have often been frustrated by the labor required to animate programmatic advertising.[12]

Traditional advertising agencies occupy an important but conflicted place in these markets. Agencies, and particularly media buying organizations, act like investment managers, deciding how to spend clients' money. Many agencies built or bought their own trading-desk systems for buying programmatic impressions. Over the last decade, companies providing advertising and media buying services, including the largest conglomerates, have branded themselves around commitments to data science and artificial intelligence. As the *Wall Street Journal* reported in 2018, "Ad giants such as WPP, Omnicom and Publicis have gone on acquisition sprees, bringing legions of information-technology experts into their ranks."[13] Caught up in the big-data frenzy of the 2010s, advertising companies (and media companies, too) poached data scientists from other industries and padded their rosters with PhDs in physics, mathematics, and neuroscience.

But agencies are also playing defense. With the profusion of automation and analytics services, other intermediaries have absorbed traditional agency roles, and some advertisers are managing more of their media planning and buying in-house.[14] A few large retailers, such as Walmart and Walgreens, even launched their own DSPs and programmatic marketplaces to take advantage of all the shopping data they collect, including the personal profiles amassed through data-extractive customer loyalty programs. Meanwhile, management and accounting consultants such as Accenture and Deloitte leverage their competencies in data analytics to chase a bigger share of the market for advertising services.[15]

The largest adtech companies play multiple roles in this commodity chain. In addition to owning highly valuable advertising inventory on its search engine and on YouTube, Google controls an integrated stack of adtech utilities, serving both the demand side (advertisers) and the supply side (publishers) of the market and serving the two sides simultaneously as an auctioneer. This multilevel control allows Google to exert enormous pressure on advertisers and publishers. Auction design, access to information, bidding speed, selective interoperability, and preferential partnerships all help manifest power asymmetries.[16] The company uses that structural power to drive and lock-in adoption of its tools. For example, advertisers that do not use Google's DSP are disadvantaged by design in certain circumstances. Because its buying and exchange platforms can share exclusive data that is unavailable to rivals, Dina Srinivasan explains, "Google-owned buying tools . . . more frequently make informed decisions about the value of inventory for sale" on Google's ad exchange.[17] Google has also leveraged its near-monopoly in the ad server market to channel publishers' inventory through its own exchange, taking a cut of the resulting trades.[18] There are even allegations that the company intervenes in its auctions to influence outcomes and prices.[19] Essentially, Google provides market-making infrastructures as for-profit services, which ad buyers and sellers depend upon to access routes of trade and to utilize the calculative devices that seem necessary for rational optimization. Other "walled garden" platforms such as Instagram, Amazon, and TikTok also contain dedicated adtech stacks.

The proliferation of adtech intermediaries has divided advertising transactions into sequences of discrete events, each of which is an opportunity for intermediaries to collect fees or gather data that can be commodified or used to render future services. Critics have noted that with so many brokers holding out their greedy hands, a diminishing portion of ad spending reaches "quality publishers."[20] Sometimes the programmatic business even aspires to internal perpetual motion. You have likely encountered advertisements that look like clickbait headlines in the "chumbox" placed by companies such as Taboola and Outbrain at the bottom of news pages. These ads, disguised as news recommendations, often redirect the user to "made for advertising" websites that exist solely to monetize this paid traffic by surrounding anemic, outrageous, or plagiarized content with heaps of ads sold by other programmatic vendors.[21] In many ways, this represents the culminating logic of a business organized to produce evidence of attention and behavior. Caitlin Petre has aptly likened digital publishers to "traffic factories."[22]

Programmatic advertising is an elaborate and opportunistic solution to the informational and administrative burdens involved in selling the American people. For all its apparent novelty, though, it revolves largely around the same interconnected operations that Oscar Gandy recognized in the discriminatory system he called the "panoptic sort"—identification, classification, and assessment.[23]

ADTECH'S PANOPTIC SORT: IDENTIFICATION, CLASSIFICATION, AND ASSESSMENT

There are many methods for identifying people on the internet. The best known is the cookie, a small text file that companies use to track web users and build behavioral profiles. Third-party cookies, left by someone other than the website itself, have been widely used in programmatic advertising to recognize individual consumers and their perceived value whenever and wherever advertising opportunities arise. But third-party cookies are being phased out, leaving many adtech companies scrambling to find new ways to persistently and uniquely identify consumers. Pixels are another tool for tracking and identification. These tiny bits of code collect detailed records of what people do on websites—where they click, what they add to shopping carts, how long they spend on a page, and so on. Facebook has placed conversion pixels on millions of websites to connect evidence of advertising exposure with evidence that the user bought an advertised product. Fingerprinting is a third technique that creates a unique identifier for a device and ostensibly the humans using that device. It is basically a distinguishing inventory of features such as time zone and language settings, hardware and software specifications, how fonts are rendered, and more.[24]

All these techniques are used to recognize people (or, more precisely, devices and browsers) across situations. But it is no small challenge to assemble the data an individual generates across devices, accounts, and online and offline shopping; associate the data correctly with an identifiable persona; and then recognize that persona when configuring or documenting marketing events. A variety of specialist companies and software systems have emerged to deal with issues of "identity resolution." Amazon Web Services, for example, markets an "identity graph" that promises a "360° view of customers." The graph, which lets marketers "establish persistent identifiers" for individuals, "is used for real-time personalization and advertising targeting for millions of users."[25] This is typical jargon for descriptions of identity solutions. The goal is to make better-than-random guesses about the value and probable behavior of users whose impressions are for sale—although, in many cases, assigning identity to web users is itself a guess generated by a probabilistic model. Identification represents a particular advantage for walled garden platforms where users log into an account.

Identification does not mean that advertisers can recognize a person by name or other human-readable markers. Adtech targets individual users not because of an interest in their personhood but based on some recognition of how a user, or even just a measurable feature the user exhibits, can be classified. A major function of adtech is sorting people into dynamically adaptive categories and populations. Adtech systems classify users, moment to moment, based on patterns or propensities that seem predictive of someone's responses to certain stimuli and their value for specific marketers. Adtech produces calculated or algorithmic publics, which may exist for only an instant as marketing opportunities appear and disappear or as a system ingests new information.[26] The inscrutability of classification decisions—the fact that people have very little insight into why, how, and with whom they are being grouped and targeted—suggests that these publics might as well be called "privates." This also gestures to the feeling that adtech classifications are dubious and malformed, hidden in proprietary databases where the sun doesn't shine.

Finally, assessment is the evaluation of individuals and populations to determine the opportunities available to them. Often it means valuation, or assigning prices or estimates of economic worth. Programmatic bidding is an interesting site of assessment. DSPs use algorithms that can be trained to discern the value of users and impressions over time. As an industry consultant explains, a DSP "starts to identify what are the characteristics of impressions that make revenue, and tries to optimize to create the most favorable ratio possible of revenue to cost. That's what DSPs do. . . . It's really sophisticated technology that optimizes on the fly to achieve your business objective." DSPs do this by first bidding randomly "in order to acquire data about which types of impressions lead to good outcomes and which ones don't," and then using the "training dataset" compiled through this random bidding to "build its predictive model," which renders assessments about whether an impression is likely to yield the desired outcome.[27] Moreover, assessments of people and advertising opportunities are adapted to different situations. For instance, someone who is classified as wealthy might fetch rich prices from advertisers selling mutual funds, whereas someone judged to be in a desperate financial situation might be highly valued by predatory companies advertising high-interest loans. Assessments about the worth of these two consumer types, and the likelihood of them clicking on these respective ads, influence what offers they see and perhaps their own sense of self.

This is the bedrock of adtech—assembling machine-readable profit projections to discriminate between good and bad bets.

PROBLEM GAMBLING

Programmatic advertising has been enormously lucrative for some companies. But it exhibits some well-known pathologies. It is notoriously incompatible with any serious definition of privacy.[28] Requests for bids on impressions can communicate personal information about a user to hundreds of companies.[29] These "bidstream" and other adtech data are sometimes recycled in situations where disclosures of sensitive details can have life-changing or life-threatening implications, such as immigration enforcement, military targeting, and persecution of minority or criminalized persons.[30] Right-wing assaults on reproductive and civil rights, for example, are intensifying the danger that the location tracking used for mobile advertising could facilitate state and extrajudicial violence.[31] The imperative to recognize the value of consumers and advertising opportunities powers a massive business in surveillance, profiling, and data brokering.

Paradoxically, by selling impressions in isolation from the surrounding media content, adtech leads to perverse decisions about advertising distribution. Advertisers' willingness to delegate bidding and ad serving to blackboxed systems has helped monetize hateful speech and disinformation,[32] and it has caused companies to advertise next to content detrimental to their brands.[33] No less perverse, when advertisers *do* dictate the types of content they want to avoid in the interest of "brand safety," they often withdraw support from news about public health, racial justice, and climate change, as well as from content creators who are deemed "risky" because they deviate from straight, white, middle-class norms.[34]

These pathologies are endemic to adtech. Programmatic advertising is emphatically not the answer to questions about how to fund democratic communication systems, how to support diverse stories and intrepid journalism, or how to improve user experiences. Programmatic advertising is a solution to logistical and epistemological problems of identification, classification, and assessment. The solution, essentially, is to make advertising markets into machinelike configurations awash in computer-processed claims and predictions, all motivated by the desire to discriminate profit potential more finely, more rapidly, and more massively—to optimize private values.

Understanding that adtech is about recognizing value helps us see that the advantages of scale and the industrial pressures toward consolidation and monopoly are systemic features. Despite the baffling number of companies under adtech's umbrella, an oligopoly rules most of this business. Google, Meta, and Amazon are head and shoulders above the rest of the field, leveraging dominant positions in search and web browsing, social media, and e-commerce. They enjoy expansive views of user behavior, easy and legally permissible access to tokens of user identity, and widespread interconnections for facilitating information flows. Platforms like Facebook record virtually everything known users do within their walls, and they exploit their integration with other websites, apps, and data brokers to connect their own observations with records of what the same users do elsewhere. Furthermore, as intermediaries that process advertising transactions, these platforms have unmatched access to marketplace information, such as bidding patterns and clearing prices.

Whatever the quality and usefulness of big platforms' big data, these companies convincingly present themselves as the actors best positioned to make credible claims about the value of each consumer and each advertising opportunity. This is why "privacy" maneuvers by Google and Apple have partly been power plays to sabotage competitors.[35] By preserving their own abilities to collect and cross-reference user data and denying competitors those privileges, these companies gain advantages in identifying, classifying, and evaluating consumers and in attributing marketing outcomes to the events they stage and sell. Marketers gravitate toward these platform companies because they promise, quite persuasively, to measure and optimize return on investment. Bigness in adtech affords truth-making and infrastructural power.

This brings us back to the chapter's central proposition: adtech is about epistemology and logistics. The term *epistemology* seems inapt for systems that may be held together by duct tape or are, in some cases, full-blown scams. But the point is that the business produces evidence, claims, and predictions that organizations and professionals act on. Adtech develops management science machines for enacting and exploiting differences in value and probability. Whether or not they work in ways that are legible to humans and outsiders matters less than the apparent fact that companies trust them to work or feel compelled to imitate competitors for fear of suffering some disadvantage. The knowledge infrastructures that support particular definitions of audience attention and behavior produce what Jérôme Bourdon and Cécile Méadel call a form of "procedural truth."[36] They organize the realities admitted into the frame of rational calculation. We often refer to advertising markets as markets for attention, but we should also recognize them as markets for particular types of evidence.

EVIDENT ATTENTION AND BEHAVIOR

Scholars have long understood that commercial media commodify audience attention.[37] How that works, basically, is that advertisers pay companies that control the means of communication to execute distribution events, delivering commercial messages to a specified number and type of potential consumers. Advertisers hope audience members will interpret those messages and, better still, act in some desired way as a result. Dallas Smythe argues that, as people decode advertisements, they participate in their own socialization into lifestyles and purchasing habits that reproduce their labor power and capitalist social relations.[38] The details of this process are complex and contradictory, but Smythe's contention is useful for recognizing how commercial media are organized to produce consumers and consumption in action. This side of the "attention economy" is about how allocations of consciousness and cultural energy make and remake social worlds. It reminds us of the prompt Harold Innis left for media studies: "Why do we attend to the things to which we attend?"[39]

At a more prosaic level, commercial media also produce consumers and consumption in information—as evidence, documents, statistics, and accounts. These operations are actually more material to the business of monetizing attention (including via fraud), and they demand an understanding of attention that is sensitive to its conditions of commodification—that is, producing and packaging observable events that can credibly be claimed to represent human cognition or behavior. Advertising transactions revolve around particular ways of defining, operationalizing, and materializing media usage. Audiences and attention, in this context, are informational products, assembled through historically specific conventions, instrumentations, and property regimes. Manufacturing these products involves making certain user activities and attributes available for measurement and maintaining industrial relations that make the extracted data intelligible and legitimate to those who would trade or act on it, including machines. Salable attention exists via the platforms and devices that manifest it and the methods of processing it into a discrete, tangible thing. What's sold is *evident attention*.

Historically, evidence of attention has meant measures of media exposure manufactured by third-party research firms that act as "neutral" intermediaries between advertisers and attention merchants. In television, for example, Nielsen gathers behavioral records from a sample population and then uses those panel data to produce "ratings" representing larger viewing publics. Ratings have functioned as an authoritative "currency" for setting prices and converting scattered and endlessly diverse audience experiences into an actionable collection of standardized marketplace facts. Panel-based audience measurement has continued on the internet,[40] but digital media also allow the companies serving advertisements to log the details of which ads they deliver to whom, as well as to observe clicks, downloads, mouse movements, and other indices of relevant behavior.[41] Advertising intermediaries have exploited the technical fact that the process of serving digital ads can automatically produce documentary records, including data about users.[42] Platforms like Facebook single-handedly execute advertising events and generate the corroborating evidence of attention or behavior. Because mistrust can arise when these audience aggregators "grade their own homework," many advertisers and publishers employ third-party verification companies, such as Integral Ad Science and DoubleVerify, to validate reported metrics and confirm that ads were loaded and viewed correctly. Despite the capacity for comprehensive record keeping, truth in digital advertising remains contestable, with companies making competing claims about ostensibly identical realities. And given that evidence of attention can be easily fabricated (often falsely, accidentally, or by internet bots), companies have their hands full sifting a torrent of evident attention to distinguish the legitimate from the illegitimate.[43]

Adtech is the name now given to "new media" infrastructures for materializing attention as a marketable commodity. "New" should be interpreted less as a matter of recency than as one of remediating relationships. Josh Lauer explains that a defining feature of "new media," in whatever historical context, is that they "produce new types of evidence, each with its own material form and truth claims."[44] Today's adtech represents the latest configuration of audience manufacture as an evidential paradigm, negotiated around the affordances of evidence-producing media, the capabilities of information systems to store and analyze evidence, and the uneven power to authorize certain evidence and enforce certain claims. "The semiotic detritus of new media, past and present," Lauer writes, "has contributed to an intensification of surveillance by introducing new forms of evidence . . . by which individuals might be identified, their motives and thoughts inferred, and future behavior predicted."[45] Adtech exemplifies this accommodation of new media and their evidence into audience manufacture.

This discussion is not meant to deny the lived practices of people making meaning with media. Rather, it highlights some specific ways that audience attention and behavior exist as objects of knowledge and commerce. The commodity audience resembles, in some sense, what information scientists call a "document." Librarian Suzanne Briet famously distinguished an antelope in the wild from one in a zoo. The caged antelope becomes a document; it can be used as a type of evidence, or index, testifying to the fact of antelope in the wild.[46] The enclosed commodity audience has a similar existence; it is documentary evidence, assembled within an organized discourse, attesting to the fact of attention or behavior. Further examination of people—where they go and what they buy—yields additional evidence about their worth and inclinations, which is drawn into the frame to valuate units of attention when they come up for sale. Advertising has always revolved around "centers of calculation," where expert communities work to collect measures of attention, behavior, and value and translate them into institutionalized claims.[47] Adtech is a bustling center of calculation and commerce for today's evident-attention economy.

Here again, the singular noun overstates adtech's uniformity. Different business models, such as online search and social media, configure evident

attention and behavior in their own ways, exploiting capacities and bottlenecks, with different implications for the organization of user experiences and the exercise of power.[48] The logic of selling the American people is
implemented with variation, activating a range of adtech tools, techniques,
and relations. Audience manufacture provides a fascinating case for looking at how documents and information technologies represent the market
to buyers and sellers and thereby structure the basis of economic thought
and action. What media researchers refer to as the "institutionally effective"
audience—evidence of attention that becomes meaningfully incorporated
into organizational routines—is constructed through a process of "framing."[49] Framing sets boundaries around what objects and relationships are
admitted into economic calculation and exchange, what is inside or outside
a market or a transaction. Framing establishes the portion of reality taken
into account and how it is "known" and represented. Because audiences and
attention can be operationalized in myriad ways, this domain is open to constant crisis, opportunity, and negotiation, although in practice it tends to
be fixed, at least temporarily, by institutional and infrastructural closures.
Perceived openings—created, for example, by new ways of generating evidence or processing transactions—trigger contests among actors trying to
cement a more profitable closure.

The historical construction of adtech illustrates these dynamics. To give
perhaps the most significant example, digital advertising has taken shape
around pressure to pay for evidence not of message exposure but of the marketing outcomes advertisers want to achieve. The long thrust of accountability has, in many ways, been about nudging the business to integrate
evidence of *audience* behavior with evidence of *consumer* behavior (see chapter 8). Not surprisingly, then, adtech is part of a bridge between marketing
and marketplaces. Digital advertisements almost always offer opportunities
for e-commerce purchases. Leading advertising platforms such as Instagram,
YouTube, TikTok, and, of course, Amazon are trying to facilitate more consumer purchasing within their own sites or apps. They are actively marketing
themselves to advertisers, influencers, and other content creators as digital
storefronts—self-contained shopping portals where users can buy whatever
they see in ads and user-generated content. Through various partnerships,
these platforms facilitate payment processing and order fulfillment in a
seamless consumer experience.[50] The idea that digital media are shoppable
media is taken for granted in advertising and cultural industries around

the globe.[51] The production of consumers is treated, here, as a technical achievement of ubiquitous connectivity to online marketplaces. As Mark Andrejevic puts it, "the entire networked environment is a space of consumption (and labor) thanks to the affordances of the digital enclosure."[52]

Enthusiasm for shoppable media reflects the dream of advertising attribution. Attribution is essentially a claim about whether or to what extent certain advertising events influenced consumers' behaviors or caused other relevant outcomes. These claims are often probabilistic, based on models that fabricate relationships between advertising events and marketing results through a mixture of observation, correlation, and estimation. And their limits get redefined by technical, legal, and cultural changes. The recent rush by Apple and Google to ensure privacy is impairing marketers' ability to track the performance of their ads in generating sales. For example, Apple's new restrictions on cross-app tracking "makes it harder to measure when an ad on one app leads to a sale on another." This is pushing platforms to further develop their own e-commerce functionality. According to a writer in *Advertising Age*, "Without the ability to follow that digital trail [across multiple websites or apps], social media companies want consumers to go from discovering products to purchasing directly within their platforms. This would allow the platforms to prove their value, and retain customer data, too."[53] Meanwhile, as media platforms are becoming shopping portals, retailers are becoming adtech companies. Limitations on third-party data sharing create opportunities for large merchants to exercise their massive customer data sets and their ability to document purchases by individual shoppers.[54]

Whatever solutions marketers and marketing platforms develop to bolster their attribution claims and adapt to new privacy requirements, the underlying motivation will be to generate credible and institutionally effective evidence of audience attention and consumer behavior. That is the grist in adtech's money mill. Importantly, though, that mill grinds a messy mash of prediction and intervention. It can be hard to tell whether apparently successful digital advertising caused sales or intercepted them. Adtech lets intermediaries stand beside the cash register and take credit for events that might have happened anyway. In other words, attribution may not measure the influence of an advertisement so much as an intermediary's success at recognizing that someone was on their way to buying something. By betting on that propensity and documenting its consummation, adtech fabricates evidence of value, justifying service fees.

CONCLUSION

In one of the best-known histories of US advertising, Stephen Fox describes a recurrent cycling between two paradigms: one that creates cultural *imagery* about brands and their users, and one that issues *claims* about products, often appealing to science and rationality.[55] With adtech—and, as I argue, with broader processes of buying, selling, distributing, and evaluating advertising—it's claims all the way down. Or, rather, it is claims, conjectures, predictions, classifications, and investment decisions, all held to account by an avalanche of statistics and documents. A key task, then, is to uncover how certain types of claims and certain types of claim makers gain or lose legitimacy. Although the regimes of knowledge and power that operate in adtech today are novel in many ways, the fact that they were recognized as offering convincing solutions and desirable opportunities for advertising must be explained through longer histories of technology (including hype and failure), political economy, and business culture. Ironically, because it is so difficult to vet many of the claims at the heart of digital advertising, the backstop for judgment often rests on a company's ability to wield science and technology within its brand image.

A focus on claims helps us position adtech within the development of informational, surveillant, and technoscientific elements in capitalist marketing. The rest of this book can be considered a contribution to what Adam Arvidsson calls a "prehistory of the panoptic sort." Arvidsson and other historians point to a shift in marketing around the 1950s, as computerization, segmentation, manufacturing capabilities, and innovations in statistical methods all facilitated a focus on specialized populations and a commitment to tracking and adapting to more flexible patterns of consumption.[56] I suggest, in particular, that this paradigm was inflected by management science, which is, in many ways, a set of techniques for enacting differences of value in the service of optimization. Adtech promises automated management science, combining logistical infrastructures with quantitative formalisms for decision making and evaluation. When we talk about programmatic advertising or the business mechanics of an advertising platform like Facebook, that is what we are talking about.

This introduction to adtech has focused on logistics and epistemology. Adtech applies techniques of identification, classification, and assessment to recognize costs and profit opportunities, to predict behavior, and to act on those predictions. My critique, therefore, starts by defining adtech as ostensible solutions to problems of distribution and management. This is different from attacks on adtech that gather around opposite poles: at one end, adtech is an irresistible manipulative force; at the other end, it is pure marketing fiction. Neither of these positions captures what's really happening: rival computer systems dueling over pennies and probabilities. The issue is not about mind control. It is about the accumulation of power by companies that administer bottlenecks and spin all sorts of evidence into claims about value and the future. Certainly, many advertisers and adtech companies dream of controlling consumer behavior. The expansion of surveillance and the proliferation of dynamic choice architectures used for "nudging" are both part of marketers' efforts to manage consumption. But, for the most part, adtech materializes elaborate designs to engineer, orchestrate, and capture value from new investment opportunities.

Management science's major contribution to adtech is formalizing the decisions involved in distributing ads and allocating investments, thus translating ad buying, selling, and serving into operations that can be acted on by computers and algorithms. A critical insight from this history is that, in chasing dreams of optimization, advertisers and their agents have been required to state explicitly what they care about and how they value different types of consumers and different types of advertising opportunities. For all its complexity and opacity, adtech exposes a simple truth: advertisers are trying to maximize private economic value, not democracy, justice, or joy in public life. Equitable politics and lively arts and culture are not the objective functions that advertising models are designed to optimize. A close look at how advertising came to emulate management science shows that counting on marketers to fund truly social media is like counting on a wish and a prayer.

This points to a key conclusion. While adtech today is undeniably grotesque, it is an embodiment of deeper problems. It is a set of strategic responses to the prospects and challenges arising from within the business of selling the American people. Over the second half of the twentieth century, a variety of actors tried using technoscience to optimize the twin goals of producing consumption and commodifying evident attention. Optimization has been such a productive ideology for advertising because it implies a definitive achievement of the best possible outcome, yet it is endlessly deferrable, open to renegotiation whenever conditions seem to change. Even as privacy rules restrict the field of play in adtech, for example, companies will promise to deliver whatever new optima are possible under the circumstances, likely leaning to an even greater extent on claims of technoscientific and logistical power.

Adtech represents a particular expression of what I call advertising's calculative evolution. Despite the breathless talk of historical rupture surrounding digital media, the discourse related to optimization in advertising has displayed remarkable consistency over time. The rest of this book reveals how the world performed by contemporary adtech resembles the futures imagined by certain advertising professionals at moments when information technologies appeared to open paths for progress, permitting better optimization—new claims of *more perfect*.

2 ADVERTISING'S CALCULATIVE EVOLUTION

Advancing skill in handling numerical data is a major facet of intellectual and material progress. Man, the marketer, is first of all Man, the counter. Whether he keeps his score on clay tablets or magnetic tapes, the perennial questions of the market place are: "How many?" and "How much?"

—Wroe Alderson and Stanley J. Shapiro

This chapter elaborates some specific arguments about transformations and continuities in the business of selling the American people. Historians remember the late 1950s as the start of a "creative revolution" that reshaped advertising in the United States and elsewhere. The agencies representing well-known brands turned their backs on the preceding "organization era," rejecting its icons of bureaucracy, rationality, science, and efficiency.[1] Without denying the significance of these developments, I argue that this moment also marks an inflection point for an equally profound trajectory in the opposite direction—a new phase in advertising's "calculative evolution." This term provides a different lens for analyzing the entanglements of marketing, media, and information technology across the second half of the twentieth century. Specifically, it helps us recognize how processes involved in market research, direct-mail marketing, and media planning and buying helped channel and amplify currents in surveillance, data processing, and behavioral and management sciences.

Social and industrial changes in the 1950s and 1960s made space for advertising to accommodate new consumer identities and lifestyles. Assembling narrower and more flexible market segments and delivering commercial messages to the right people called for systematic planning and decision making.[2] Agencies retreated from scientific approaches to preparing copy and art, but they moved decisively in the other direction in determining how to allocate advertisers' money. Even as advertising's place in popular culture became aligned with hip creativity, certain parts of the business absorbed the rationalizing pressures of Cold War politics and capitalist technoscience. Frequent and not-always-coherent efforts to account for more aspects of consumer and audience behavior empowered technicians and managers who promised to optimize objective metrics of efficiency and success. Operations that focused on distributing commercial messages, evaluating advertising opportunities, and measuring outcomes all hinged on methods of identifying, classifying, and appraising consumers. The intensification of a quantitative logic in audience construction and advertising distribution helped tune the heartbeat of the whole marketing system to the rhythms of rational calculation, in concert with the droning roar of computerization.

I do not mean for the term "evolution" to be taken too literally, and I certainly do not mean to imply normative improvement. My motivation is partly rhetorical, to draw a contrast with the creative revolution. That shift began at the same moment advertisers and agencies ramped up computerization and invested in management sciences. Over time, operations associated with data processing and optimization became magnets for the ambitions and eventual authority of calculative experts, lately nicknamed "quants" or "math men and women."[3]

Another reason for using the term is to recover the long histories of consumer surveillance, data mining, algorithmically aided decision making, and tendencies toward technical and mathematical expertise in advertising and marketing. I want to distinguish my claims—that pressures toward surveillance capitalism have been central to the businesses that commodify audiences and organize communications media as extensions of the marketing system—from claims that companies like Google recently pioneered a disruptive new mode of accumulation.[4] There have been undeniable changes in advertising associated with the internet, mobile devices, social media, and artificial intelligence, but those changes are rooted in a deeper tangle of capitalism, science, management, business cultures, and information technologies. Furthermore, those changes would not have been practicable without prior developments in these industries' underlying infrastructures and institutions, including logistical and administrative systems for organizing flows and contested hierarchies of knowledge and legitimacy. Surveillance capitalism vibrates with the echoes of earlier calls to optimize advertising and consumer behavior. To put it a bit too bluntly, it may be that advertising and marketing revolutionized the internet, not the other way around.

The calculative evolution, then, is meant as a critical heuristic for rethinking continuity and change. The next few chapters focus on a midcentury inflection point in that evolution. This chapter sets the scene by describing the historical roots of calculative cultures in advertising. Some of the attributes that seem distinctively digital are, we might say, more like expressions of the DNA in selling the American people.

ADVERTISING AS A FIELD OF KNOWLEDGE PRODUCTION

Modern advertising has always had an appetite for science. Many practitioners and commentators have believed, or found it convenient to suggest, that advertising follows discernible regularities, with success measured against objective evidence. The golden criterion is usually sales. Proof is elusive, but for more than a century, advertising professionals have promised to tame the uncertainties of commerce with expert knowledge and technique. These distinctive capacities have been central to status and authority in the advertising business.[5]

Historians have tended to focus on how debates about science in advertising manifest in competing approaches to copywriting and persuasion. Close attention to fashions in creative production follows from the supposition that advertising is about producing ads. Certainly, advertisers want effective or meaningful ads, and plenty of research arranged by advertisers and their agencies has been dedicated to that goal. But advertising is not only about creating specific types of commercial messaging. As Joseph Turow points out, "lavishing attention on what trade parlance calls the 'creative' side of the business leaves out essential aspects of advertising's social role."[6] The broader advertising system produces populations, processes and circulates business information, distributes commercial messages, and evaluates success. "Advertising," here, refers not just to the creation of advertisements but to a vast network of relationships involving merchants, manufacturers, corporate management, marketing agencies, media companies, administrative researchers, and information and technology providers attempting to accelerate and track the circulation of commodities. Advertising is a strand of capitalist technoscience.

There is an understandable urge to qualify advertising "science" and to discount it from histories of science and technology, despite the industry's long record of organized commercial research. In general, advertising has helped itself to whatever theories and techniques seemed useful for dazzling clients and disarming superiors.[7] Researchers may be less committed to finding the best answers than to finding the answers best suited to their company's needs. According to one possibly embellished anecdote, when an operations researcher devised a mathematical model that suggested advertising agencies ought to spend much more money on preparing and testing their creative work, "a leading New York ad agency offered a 'lifetime supply of martinis' to the analyst who could refute these results."[8]

To be sure, advertising research itself is often a form of advertising or negotiation. But an applied technoscience need not satisfy outsiders' expectations to have power within organizations and industrial fields. Advertising has taken a pragmatic stance toward knowledge production, recognizing that "things *perceived* as real are real in their consequences," to borrow a phrase from Geoffrey Bowker and Susan Leigh Star.[9] Parts of the advertising business embraced the legitimizing force of science as a discourse for organizing people and processes, and they built institutions around information and truth claims that could be mobilized to some advantage. Agencies leveraged specialized techniques and expertise to help position themselves as intermediaries among advertisers, media businesses, and consumers. The uncertainties, rivalries, and misalignments of trust among these actors have often accentuated the reliance on numbers and objectivity rituals.[10] The business of audience measurement, whose "ratings" mediate advertising transactions and other industrial activities, provides a vivid example of using strategic truth-making conventions to stabilize contestable realities.

Whether or not we think this should count as science, advertising was and is involved in the production of subjects and knowledge—about markets, consumers, audiences, behavior, and influence. That knowledge, however selective and strategic, structures all sorts of consequential choices and relationships. Commercial knowledge practices are threaded throughout modern political economies and material cultures. Major parts of the advertising industry have used science and technology to produce consumers, and a consumer society, in action and in information.

Advertising's promiscuous relationship with science is, in fact, crucial to this story. It has allowed distinct, divergent, and even highly dubious paradigms to coexist and contribute to this calculative evolution. For example, twenty-first-century machine learning differs in important ways from postwar operations research, yet I argue in subsequent chapters that operations research and other management sciences, despite their early failures, made today's "algorithmic episteme" possible by intensifying investments in information infrastructures (including data collection, algorithms, and quantitative or technological expertise) and socializing marketing professionals in the languages of probability and optimization. These disciplines helped produce a culture of calculation in which people were prepared or even compelled to accept decisions that had been made using computers and complex mathematics, despite often not knowing how they worked. This history of advertising technoscience reveals a not-quite-coherent but still-productive rationalizing project. Its nascent tendencies were already there at the outset of modern advertising, and a dramatic inflection point in the middle of the twentieth century defined the idiom for data-driven digital advertising. The result has been less a cumulative body of truth than the iteration and empowerment of techniques and professional habits of optimum seeking.

INFORMATION, TECHNOLOGY, AND MANAGEMENT

Advertising optimization has deep historical roots. Industrial capitalists adopted systematic accounting and management techniques developed to control enslaved people on plantations and to administer empires and adapted them to enterprises requiring less violent means of command.[11] Tendencies toward calculation were flourishing in parts of US business by the turn of the twentieth century. Confronting a "crisis of distribution,"[12] modern advertising and mass marketing were part of a broader pattern of leaning on information and communication resources for the efficient management of expanding operations. In particular, they tried to manage flows of commodities and practices of consumption to keep pace with the speed and scale of productive output.[13] The tethering of consumer-demand management to an attention economy helped stimulate and shape the mass distribution of facts and stories about politics, markets, crime, conflict, scandal, and whatever else might boost the circulation of print media. As Harold Innis puts it, "the exploitation of human curiosity and its interest in news by advertisers anxious to dispose of their products created efficient channels for the spread of information."[14] Likewise, the "rapid and extensive dissemination of information" within organizations and markets was essential for synchronizing labor, capital, and commodities to the technological and geographic intensities of industrial capitalism.[15] "When we think of the Industrial Revolution," Dan Schiller explains, "we may picture grimy smokestacks, a clattering din of machinery, and 'unskilled' factory 'hands.' Even at its outset, however, the process of industrialization also gripped information generation, processing, and management. Corporate information processing bulked up to process the sales orders, support the accounting and financial initiatives, and—beginning in the late nineteenth century—develop the systematic marketing and technology research plans that were all predicates of national capital."[16]

Bureaucratic restructuring saw large corporations delegate more decision-making responsibilities to middle management and outside experts hired to apply specialized know-how to firms' marketing needs.[17] Across the second half of the nineteenth century, advertising agencies transitioned from a sales force for newspapers, to independent brokers buying space wholesale from publishers and selling retail to advertisers, and into hired representatives of the advertiser, eventually providing a suite of services and being paid in the form of commissions on media spending. This last metamorphosis spawned the "full-service" agencies that dominated throughout the twentieth century. They decided where to place ads and purchased access to audiences (functions called media planning and buying), prepared art and copy for clients' campaigns, conducted market research, and later produced broadcast programming. N. W. Ayer, J. Walter Thompson, Lord & Thomas, and others got the ball rolling for Batten, Barton, Durstine & Osborn (BBDO), Young & Rubicam, McCann Erickson, Leo Burnett, and Ted Bates to manage the business of influence in an electronic age.

Advertising agencies and the newspapers and periodicals supported by advertising tried to bolster their legitimacy by marketing themselves and defending their strategies and choices in vocabularies drawn from systems, science, and efficiency—all considered agents and indices of progress. Both these businesses would claim special insights into the public's tastes and, at times, present themselves as modernizing forces in American culture and economy. A mere commitment to evidence-based judgments, whatever the quality of the evidence or judgment, fit with cultural dispositions toward discipline and rationality in business. Particularly from the 1890s onward, many practitioners and commentators invested in the idea that the advertising trade could cement its economic and cultural power by using scientific techniques to analyze and manage the details of markets and consumption.[18]

Advertising agents pinned some of their earliest claims of expertise on their skills at selecting advertising placements and buying access to audiences of consumers. Their authority derived from their experiences and relationships with publishers. Buying frequently and in bulk from many publishers, agents professed to have more knowledge and bargaining leverage than the advertisers they represented. For starters, they simply knew better than manufacturers and retailers what publications were available for advertising. Enterprising agents began organizing information about newspaper circulations and advertising rates by the mid-nineteenth century. Volney Palmer, serving as a sales agent for newspapers, shared this information with advertisers; George Rowell, an independent space broker and later the founder of the trade magazine *Printers' Ink*, published and sold collections of such facts as media directories.[19] Buyers could consult these resources as they sought to place ads and negotiate prices, and publishers could wield them for promotional purposes and to justify their rates—assuming the reported figures were flattering. These information digests also made it clear that advertisers and publishers were participating in a market for audiences. Describing *Rowell's American Newspaper Directory*, Zoe Sherman writes, "It reads as a catalog of the size and character of the readership of each newspaper."[20]

As advertisers became the primary clients of media buying services, the well-informed agent's pitch often hinged on finding the most cost-efficient placements. Rowell followed this tack, advertising his own agency in 1896 by promising a good return on investment: "Successful advertising means getting back more money than is paid out. . . . In this agency of ours, the constant endeavor is to make the cost smaller and smaller, and the results larger and larger. We have succeeded better than others simply because we know how to succeed." Among its sources of success, Rowell's company boasted of efficient media selection "by using the best media and rejecting the bad."[21] Some agents promised access to especially valuable audiences. By the 1870s, J. Walter Thompson (JWT) had a near monopoly over magazine advertising.[22] In 1887 JWT distributed a "catalog" of magazines, instructing advertisers in the benefits of using magazine placements to discriminate by class. By targeting prestigious titles, advertisers would gain "entrance to the *better* class of homes," populated by people of "taste" and "means," while avoiding the "ignorant classes [who] have no inclination to spend money" and "the poor [who] have none to spend."[23]

JWT was leading the pack here. At least until the 1890s, plenty of agents simply tried to procure the most space and circulation each dollar could buy (or, sometimes, they targeted publishers that paid higher-than-normal commissions to buyers). Nevertheless, it is clear that people in the business recognized that populations could be classified by their profit potential to advertisers, that different publications organized these populations into different audiences, and that strategic media buying could secure efficient access to desirable customers. These were relatively crude classifications, but they implied that inclusion or exclusion in this fast-growing part of public life—whether one counted in the American culture represented in popular media—depended on assessments of one's value as a consumer. A practical knowledge of class distinctions was an important part of the expertise modern agents sold to clients.[24]

Leading agencies conducted research and stockpiled information, responding in part to the demands of large advertisers that "concentrated on technique and its effectiveness."[25] By 1879, N. W. Ayer & Sons, a pioneer of the modern agency form, began assembling figures about its clients' industries and markets. In the 1880s it solicited new business by first studying a prospective client's business and then "calculating how best to improve on its current marketing by more efficiently placing periodical advertisements."[26] Publishers of mass-circulation periodicals also helped cement the production of consumer data as an institutional mainstay in ad-supported media. Urged on by demands for information about the readers accessible through these magazines, publishers sized up the lifestyles of lucrative audiences and discounted less prosperous groups (e.g., Black Americans, ethnic immigrants, rural poor), unless perceived vulnerabilities made otherwise undesirable people attractive targets for exploitative marketing.[27] Publishers eagerly collected names and addresses, as well, appreciating these mailing lists as valuable assets.[28]

At the behest of manufacturers and merchandisers, advertising and media industries were ramping up the production of consumers. This meant not only nurturing new tastes and buying habits but also producing knowledge to make consumers legible within contemporary systems of management and capital accumulation. As publishers, agencies, and advertisers examined US markets and audiences in search of profitable and often self-serving insights, their activities helped construct certain ideas about American consumers and their lived realities and operationalize them for business purposes. In alignment with cultural assumptions and the marketing priorities that motivated this research, they codified the ideal consumer as a white, middle-class buyer.[29] Expert knowledge of consuming subjects and the means of finding and managing them was a main credential of the modern advertising industry. Indeed, the industry was actively "making up people,"[30] defining and producing subjects for business to observe, record, and act on.[31]

"THE MATHEMATICS OF ADVERTISING"

A sensibility that drew on both accounting and engineering was gaining currency in the advertising business by the early 1900s. Earnest Elmo Calkins and Ralph Holden describe the scene in their canonic 1905 textbook *Modern Advertising*: "The present-day tendency on the part of experienced advertisers is to get at the facts—to reduce the art of advertising to a science—to develop what may be called the mathematics of advertising." Just as "the successful architect is capable of calculating the breaking strain of an iron beam," the successful advertising campaign—meaning one that has proved to pay off in sales—can be worked out through the "statistics of advertising."[32]

Such a mathematics implied suitable knowledge infrastructures. In a 1909 manual prescribing systematic principles for the organization of advertising and sales departments, business theorist James Bray Griffith acknowledged that the "advertising manager collects a vast amount of information for future use." To take advantage of this information, "the advertising man must have a system of records which will show at all times what he is doing, what results he is getting, exactly where he stands."[33] Businesses used various physical and intellectual devices to manage paperwork and the relationships and activities those documents represented or facilitated.[34] This included storage equipment and filing and indexing schemes for keeping track of all sorts of resources that an advertising manager needed to consult or retrieve. Direct-mail advertisers were particularly interested in office technologies that helped them quickly produce and dispatch customized print materials. Cards containing updatable information about customers, machines for addressing envelopes, and a low-wage, feminized clerical workforce were pillars of the direct-mail business in the early twentieth century.[35] That business helped build and host some of "the largest repositories of personal data in existence" at the time.[36] Advertisers and agencies also employed staff to verify that publishers distributed ads in accordance with contracted orders and to reconcile any apparent discrepancies—jobs that persisted throughout broadcasting and into the ostensibly automated environment of programmatic advertising.[37]

A prime motivation for assembling this information, as Griffith saw it, was to support "intelligent" decision making and to calculate return on investment: "A necessity to the advertising department is an efficient system of checking and recording returns from advertising. Without a system on which he can depend, the advertising manager is spending his employer's money blindly—he does not know what he is *getting* for it."[38] Adherents to this advice arranged their procedures to impose documentary order—to make decisions by the books. In time, more and more businesses used punch cards to record data and machines to sort and tabulate them.

Some of Griffith's contemporaries said it was not enough for advertising management to be systematic, as even wasteful activities could be systematized. Business writer Herbert N. Casson advocated a scientific approach to efficiency. In 1911 he suggested that Taylorism's scientific management of workers and work processes should be applied to the sphere of circulation: "One by one, almost every activity of man is being analyzed and organized and uplifted into a science. . . . And now the next great step, in the general swing from metaphysics to science, is to apply the principles of Efficiency to the selling and advertising of goods. What has worked so well in the acquisition of knowledge and in the production of commodities may work just as well in the distribution of those commodities."[39] In defending his proposals from critics who might argue that advertising was misfit for scientific management, that its effects were inaccessible to observers and forecasters, Casson took inventory of businesses with evident capacities for knowing and analyzing consumer populations:

> To say that the public is an uncertain quantity and cannot be measured is absurd. The insurance actuary measures the public. . . . And his knowledge is so accurate that hundreds of millions of dollars are staked upon his calculations. . . . The experts of the telephone companies measure the public. . . . Railway and steamship companies measure the public. . . . Newspapers measure the public best of all, perhaps. . . . The magazines, too, measure the public. Their very life depends upon these measurements. . . . Immense businesses are based upon the fact that the activities of the nation as a whole can be foreseen. Just as there are to-day

actuaries who predict public health, so there may be actuaries who will predict public opinion in its relation to the sale of goods.[40]

We might gather from Casson's defensive tone that not everyone believed advertising could be formalized in these ways. A study of *Printers' Ink* concludes that in the two decades after its founding in 1888, most writers "agreed that advertising was not and could not become a science, that, in fact, it was too much a matter of chance even to make it worthwhile to test statistically the results of a particular advertisement."[41] But that same study observes a reversal in expert opinion around 1910, as "the language and methods of applied science became the order of the day." Advertisers were especially keen to discover "the precise reactions to advertising, the actual preferences of the public, the measurable habits of customers."[42] As historian Pamela Walker Laird puts it, advertising specialists in the early twentieth century claimed to lean less on their own intuitions and more on an empirical "sensitivity" to what the buying public wanted, thereby "increasing the probabilities of appealing to consumers."[43]

This desire to know more about consumers and advertising effects was evident in agency practice. Leading firms were opening feedback channels and building organizational capacities to account for advertising's influence on sales. By 1900, the Lord & Thomas agency had instituted a "Record of Results" department that reported weekly to clients on the performance of their campaigns, as measured by replies to mail-order advertisements and correlations with retail sales.[44] With a staff of eight people "tabulating and filing returns" related to more than six hundred clients and four thousand magazines and newspapers in 1906, the department provided "a steady flow of hard data with which to pick media and impress clients."[45] Maintaining this information flow as a regular routine, the agency presented its media buying services as systematic, efficient, and oriented to the advertiser's return on investment. As Stephen Fox explains, "Ad placement, traditionally dependent on a medium's reputation and a sixth sense of the agent's, thus acquired a more rigorous procedure."[46]

To gauge the reach and impact of their print advertisements, advertisers and their agents used schemes to solicit feedback from readers, such as issuing coupons that could be redeemed at stores or mailed back to manufacturers. Advertisers tried to attribute results to specific placements by marking coupons with a "key" that identified the publication and issue. These returns not only supplied evidence of advertising exposure but also allowed some calculation of the cost of acquiring customers and obtaining sales.[47] Historian Stefan Schwarzkopf describes keyed advertising as the birth of "measuring return on investment (ROI) through market research."[48] Sometimes print ads also asked respondents to submit information about themselves or people they knew, which could be used for additional advertising or sold to data brokers.[49] At least a few agencies used direct mail to target people who responded to clients' ads, a form of follow-up advertising that works on the same logic as digital retargeting, where people who have signaled something about their interests or needs are singled out as prospects worthy of further investment.[50]

These designs were far from perfect, but they showcase advertisers' and agencies' desire to increase their stores of knowledge, to scaffold strategic decision making, and to evaluate the efficiency of their efforts. Information assembled from dedicated and seemingly systematic feedback systems was becoming the foundation of credible judgments about consumers and audiences—who they were, what they wanted, and how to reach and persuade them. Advertising professionals were only selectively rigorous, of course, and they continued to draw on their own intuitions and assumptions. But at least outwardly, advertising managers and intermediaries projected a commitment to knowing and managing consumers scientifically. Consumers would exist not only in the musings of cultural diviners who created art and copy but also on paper, in the accounts and statistical records maintained by advertisers, agencies, and other companies or bureaus. Recourse to these datafied consumers and the means of making sense of them became a calling card of advertising specialists.

MARKET RESEARCH INSTITUTIONALIZED

American business committed even more energy to knowing consumers in the 1910s, as market research matured into its own service sector. Dedicated research divisions became permanent fixtures at many agencies, publishers, and advertisers, and independent vendors of information or research services emerged.[51] According to a prominent operations researcher, this market research, and the marketing science that descended from it, was motivated by the desire "to reduce uncertainty in the marketing process."[52] The "fundamental intent" of early efforts to "categorize individuals by means of demographic-type variables," he notes, "was to define and identify *potential* customers in the population. As advertising became a force in the marketing process, the problem of matching media to marketing prospects gave great impetus to these efforts."[53]

The Curtis Publishing Company, best known for the *Saturday Evening Post* and *Ladies' Home Journal,* was especially prolific when it came to making American consumers a matter of record. The company established in 1911 what marketing consultant Wroe Alderson calls the "first full-fledged market research organization in the United States."[54] The department thrived under the direction of Charles Coolidge Parlin, "one of the earliest pioneers in the infant science of marketing."[55] Starting in 1913, Parlin mapped consumption patterns in an *Encyclopedia of Cities.* Curtis Publishing went on to produce many volumes, reportedly used by hundreds of businesses, that rated states and counties across the United States according to their sales potential. In the 1920s it even divided large cities into areas of distinctive "residential character," such as those with "many fine homes" or "mostly Jews, Negros and Foreigners"—classifications that indexed racial and ethnic redlining.[56] Much of this early market research involved surveys, interviews, and scavenging from the databases amassed by government agencies. The Commercial Research Division at Curtis sometimes scavenged public collections quite literally, analyzing trash discarded by a sample of Philadelphia residents from each income quartile. The company also audited household pantries to examine the buying behaviors of consumers across sixteen states.[57] Parlin's work for Curtis introduced new techniques for measuring the distribution of goods, the purchasing power and habits of different classes of consumers, and the circulation of magazines.

As advertising agencies touted commitments to evidence-based decision making, their appetites for information continued to grow. But even as information poured in, the people preparing advertisements and planning campaigns did not always let new facts displace their existing presumptions about who consumers were, what they wanted, and how best to influence them.[58] Still, demands for more data made for good business and good theater. In 1914 E. G. Platt of JWT composed a litany in *Printers' Ink,* urging publishers to furnish agencies with detailed information about circulation, market conditions, and readers—how they lived and what they bought. Among several pages of requests, he asked for "facts regarding the influence of your publication to produce sales."[59] JWT, one of oldest and most influential agencies, was already taking matters into its own hands. By 1912, it was mining US census data for statistics that might benefit clients. Within a few years, the agency positioned scientific commitments and competencies as a cornerstone of its own public image and of the advertising profession.

JWT established a Statistical and Investigation Department in 1915 and incorporated new mechanisms for collecting and analyzing data into routine work flows. The next year the firm was bought by a group that included Stanley Resor, the first university graduate to lead an agency. Resor was an apostle for research and fact-based judgment. He saw science as a means of improving efficiency in marketing and securing professional status for ad agencies. He also believed that discernible "laws" governed human behavior. As early as 1919 he embraced the role that behavioral economists have come to call "choice architect," asserting, as Peggy Kreshel reports, that "the work of advertising was 'to control and guide the conditions that lead people to make decisions.'"[60] In 1921 he hired John B. Watson, a former president of the American Psychological Association whose program of behaviorism treated psychology as a disciplined science of prediction and control. Watson became a spokesman for JWT and Resor's vision of advertising, addressing audiences of business leaders and trade associations and "spread[ing] the doctrine of science with an almost evangelical fervor."[61]

These convictions resonated with a business climate enrapt by efficiency, science, and control and confronting challenges in distribution and marketing. Resor recognized that compelling demonstrations of advertising's power to manage consumption—akin to industrial engineers' management of production—would confer unique authority onto agencies. By the mid-1920s, he had assembled a staff that included 105 college graduates, five with PhDs.[62] Among them was Paul Cherington, a disciple of scientific management who left Harvard Business School to become JWT's director of research. There he conducted countless consumer surveys, tabulated data, and, as Fox puts it, "came up with astringently quantified answers to the imponderables of advertising."[63] Cherington boasted in 1924, "There is scarcely an advertising campaign, or a selling plan of first magnitude now operating, which has not back of it at some stage a study of the market to determine not only the number of actual or potential customers, but to get as good an idea as is possible of who they are, what their economic status is, what their buying habits and practices are, and what control their purchases of the goods to be sold." Attention to data had elevated advertising beyond "mere 'puffing,'" he said, and excised "many wasteful elements of chance" from marketing processes.[64]

For decades, Resor and company evangelized advertising as a profession administered by technically skilled experts, a powerful lever for managing consumption, and an indispensable institution in American business. Circulating through academic journals, trade associations, and other spaces of knowledge making and community formation, this discourse was ornamented with the trappings of scientific rigor and professional credibility. Advertising specialists became hailed as consumption "engineers," fine-tuning the machinery of industrialism.[65] The marketing discipline benefited further from the New Deal's legitimation of quantitative social science, statistical records, and calculated intervention in economic matters. Blaming the economic depression on marketing problems (which had marketing solutions), the discipline aligned its special expertise with the pursuit of national prosperity.[66] Even as many advertising leaders actively opposed the welfare state, they seized on chances to promote themselves as indispensable to the workings of markets and capitalist democracy. In a mission statement circulated to JWT staff in 1955, the world's leading agency called advertising "an essential function in a modern, dynamic, industrial society," necessary to the maintenance and spread of "economic well-being." Advertising, the statement said, was thus an "engineering job."[67] Whether the practice of science or the public performance of scientism registered more forcefully, JWT helped construct an agenda and a professional culture that have remained with advertising. Resor and others solidified the collection of market and consumer data as "a professional characteristic of the advertising industry."[68]

The information management systems used throughout American business provided a material complement to this inclination toward prediction and control. During the first half of the twentieth century, many marketers renovated their knowledge infrastructures in ways that afforded a fuller accounting of commercial realities. A 1932 textbook describes some of the promising advantages of "electric sales records," particularly for analyzing consumer habits: "The most highly developed technique for measuring buying behavior is that made possible by the electric sorting and tabulating machines. These ingenious devices have made it feasible to record and classify the behavior of the buying public as well as the behavior of those who serve that public, on a scale heretofore impracticable." These technologies not only supported expansive record keeping but also enabled businesses to deduce facts and patterns that were otherwise inaccessible. The textbook, which has a foreword by John B. Watson, states that the application of these techniques "exemplifies the behavioristic psychological method in almost its ideal form. It is the quantitative study and analysis of human behavior in the *nth* degree."[69]

Extracting novel insights from larger records, companies made consumers, markets, employees, and business operations legible in new ways. Department stores, for example, created punch cards representing each purchase, which were then sorted and tabulated using machines. This allowed an understanding of consumers grounded in "what the 'behavior register' actually records," enhancing businesses' capacities to cope with "the uncertainties of [customers'] future behaviors."[70] These were the big-data systems of the age, providing the straw that quantitative researchers tried to spin into gold. When JWT wanted statistical experts to discern correlations between personality types and customer loyalty, the agency offered up eight million IBM punch cards representing a full year (1956–1957) of grocery purchases by some three thousand households.[71] Keeping track of consumption had become an industrial habit.

MAKING AN IMPRESSION: PRODUCING EVIDENCE OF ATTENTION

Data collection not only let firms refine their operations or market new services to clients; it also supported durable relationships among trading partners. Brands and retailers had enlisted mass media to promote their businesses and to instruct the public about consumption habits. This model of commercial sponsorship generated demand for information about how people encountered advertisers' messages and how exposure to those messages correlated with shopping or other relevant activities. It also called for research institutions to maintain trust in the face of conflicting interests. For example, advertisers, agencies, and publishers created the Audit Bureau of Circulations (ABC) in 1914 to provide authoritative information about newspaper readership. This "represented a great milestone," as one observer later put it, since ABC replaced the "nebulous" and competing claims of rivals with a tangible criterion of circulation.[72] Recourse to a reliable and shared set of facts helped buyers feel confident that they got what they paid for. Certifying a form of "truth" for multiple parties to act on, the companies that measured media audiences built accounting (and, by omission, discounting) infrastructures. They ordered reality to support routine transactions and rational management.

Research on audiences, and how advertisements influenced their attitudes and behaviors, proliferated in the 1930s and 1940s. The production of scientific claims about public opinions and habits was a growth industry powered by prominent researchers such as Frank Stanton, Paul Felix Lazarsfeld, George Gallup, and Herta Herzog, as well as new methods of population sampling and statistical analysis. As historian Sarah Igo explains, some of the leading social-science surveyors, who were central to constructing the idea of an American mass public, "were first and foremost market researchers, devoted to the science of improving corporate profitability through carefully crafted advertising campaigns and public relations stratagems."[73] According to Todd Gitlin, the mainstream of mass communication research emerged from a "marketing orientation" and the commercial need for audience measurement.[74]

The need to know about advertising exposure prompted bursts of ingenuity. Starting in the 1930s, national magazines invested in what one researcher later called "an avalanche of audience studies."[75] Distrust in self-reports from surveys and interviews fueled interest in obtaining more direct behavioral measures. For example, researchers at *Time* magazine tried to observe "page traffic," the reader's movement from page to page, by collecting used copies and analyzing the fingerprints left on them.[76] This held the additional promise of being a personal index, unique to each reader, but overall, this method had minimal significance.[77] A textbook by professors at New York University and Northwestern University, both of whom were consultants to top advertising agencies, was more impressed with a method used by the *Saturday Evening Post* and *Look* in the late 1950s. In experiments arranged by Alfred Politz Research, the magazines applied a tiny amount of glue between pages, collected a sample of the issues in circulation, and then counted the glue seals that had been broken. Strategic placement of the glue allowed inferences to be made about whether the reader likely looked at whole pages and ads—a precursor to online viewability metrics.[78]

Politz's company also used photography to generate evidence of attention for the National Association of Transportation Advertising in Philadelphia (figure 2.1). Two cameras, controlled by timers that triggered photo bursts at scheduled intervals, were installed in the window of a city bus, where they captured images covering a 120-degree range. "Only persons clearly photographed and with both eyes visible from the bus were counted—on the assumption that they were most likely to see the advertisement."[79] This

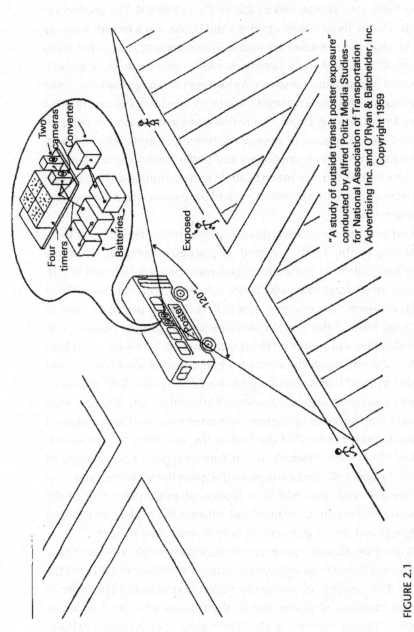

FIGURE 2.1

Researchers produced visual evidence of outdoor advertising exposure by placing automated cameras on a bus displaying poster ads. This is a colorful demonstration that an advertising impression is less about the impact of a message on a consumer and more about the impression a consumer makes on a measurement instrument. In this case, the light bouncing off pedestrians' bodies was captured by a lens and recorded on film. (*Source:* Darrell Blaine Lucas and Steuart Henderson Britt, *Measuring Advertising Effectiveness* [New York: McGraw-Hill, 1963], 281)

"A study of outside transit poster exposure"
conducted by Alfred Politz Media Studies—
for National Association of Transportation
Advertising Inc. and O'Ryan & Batchelder, Inc.
Copyright 1959

literally set the frame for what was taken into account, in terms of both territory and the definition of attention (two eyes visible).

Fingerprinting, glue seals, and mounted cameras are all methods of rendering attention in some evident, tangible form—what we might think of as making an impression. The term "impression" is usually presumed to imply an effect on a consumer; perhaps more aptly, the selling of attention requires the opposite—that the consumer makes an impression on a measurement instrument.

Pressure to create evidence of audience attention deepened with the advent of broadcasting. Reception of radiated frequencies was a fugitive phenomenon that was even harder to estimate than the circulation of print media. The audience measurement business cycled through a variety of methods to apprehend listening or viewing behavior, including telephone surveys, paper diaries, and meters that passively recorded the tuning of radio or TV receivers. Television's production costs and engineered economics of scarcity amplified the demand for data to justify advertisers' enormous expenditures and the rents extracted by networks and stations. The competition to supply that data sometimes tended toward science fiction spectacles. To dramatize its trial of a real-time audience measurement service in 1957, the American Research Bureau installed at its headquarters a giant electronic display board that visualized live network ratings at ninety-second intervals, "a kind of cross between a pinball machine and a Univac." *Billboard* reported that these blinking lights put creative workers' powers of fascination under a microscope, since a sponsor could see if viewers abandoned its show during a lousy comedy act or a dull commercial message. "Now we'll be able to watch ourselves lose clients," an agency research director joked.[80] Reflecting on the state of audience measurement in general, one advertising executive said, "Television produced an explosion of information which has required a whole new breed of media men. It generated new universes of research, and each medium produced a blizzard of research to counter the deadly thrust."[81]

This blizzard was bewildering. Different research methods produced findings that were hard to reconcile, and stakeholders could gravitate to evidence that supported their interests. Even as advertisers wanted to know about marketplace realities, they wanted to stabilize those realities for routine decisions and actions. Irresolvable knowledge claims were not conducive to that. Most ad-supported media settled around a monopolist provider of audience measurements, owing in part to the convenience and economy of operating from one official representation of reality. As many analyses of media ratings have demonstrated, audience measurement companies, acting as arbiters of truth, have provided a powerful infrastructural service for supporting the commodification and exchange of a product (i.e., attention) that exists, as an institutional fact, through the instruments, compromises, and transactions that materialize it in specific ways.[82]

From the outset, then, audience manufacture and efforts to manage consumption have generated strategies and instruments for documenting, analyzing, predicting, and potentially influencing the behaviors of individuals and groups. Referring to an emergent information industry that grew concurrently with the commercialization of broadcasting, historian Karen Buzzard writes, "The single largest category of data provided by the information industry was research on the marketing of products, including the use of the various advertising media to accelerate sales."[83] That firms such as CBS, Curtis Publishing, and Nielsen attempted to audit households' pantries so that advertisers could infer how their expenditures contributed to sales indicates the extent to which market researchers were penetrating the domestic sphere in the first half of the twentieth century. Failure to maintain and replicate these services at a viable cost also indicates the difficulty of taming such complex phenomena.

INFERENCES OF INTENTION

One last venture into complexity is worth noting here. Motivation research (MR) was a cultural phenomenon in the 1950s. It was also something of a scandal, following Vance Packard's sensational indictment of it in *The Hidden Persuaders*.[84] MR developed from social-science approaches to analyzing marketplace choices and was attached most famously to Ernest Dichter.[85] It claimed to discern the motivations behind consumer behaviors and attitudes. In contrast to Watson's behaviorism, which dismissed people's explanations of their conduct and focused instead on observed actions, MR's methodological centerpiece, the depth interview, was designed to put consumers' choices, loyalties, habits, and tendencies into words. MR thereby brought consciousness, and even *sub*consciousness, into view. Researchers exercised discretion about whether and how much to trust what subjects said, and they made choices about which details to register and emphasize. As such, MR was an idiosyncratic science. Interviewers and interpreters were crucial parts of the instrumentation both for producing data and for analyzing it.

By investigating consumer behavior in depth, MR expanded the frame of reality, even if the answers it produced could be slyly reductive. Its insights often pointed toward themes typical of advertising's creative revolution, and agencies used MR to guide creative strategies. But this mode of knowing was not easily accommodated to routine processes, especially when high-speed decision making buckled under so much individual texture. A 1961 textbook points out that MR's mode of information gathering made it "both costly and also difficult to reduce to quantitative tabulations."[86] MR might help interpret consumer behavior, but a retailer's ability to act on its insights was limited by the merchant's access to individuals' hidden inclinations. As one marketing consultant explains, "The dealer has no way of discovering the personality traits of an individual consumer, let alone the consumer's deep-seated psychic motivations. For this reason, although knowing that the person now entering the store had latent tendencies toward homosexuality would be really quite useful for marketing purposes, this information is hardly ever available."[87] This is, in some sense, precisely the problem online tracking and profiling try to solve: they make consumers' propensities apparent and actionable to marketers.

The consultant just quoted was Michael Halbert. He made these remarks in a research paper presented to the American Marketing Association in 1955. Apart from its cynicism, the paper was notable in that Halbert explicitly built a bridge between MR and the newest breakthrough scientific phenomenon: operations research. MR deviated in many ways from the quantitative thrust described in this chapter, presenting its unique sensitivities as a relative strength over the dehumanized statistics of other social sciences. Not surprisingly, that invited a strong response from advertising professionals who saw idiosyncrasies as impediments to scientific and systematic management. With operations research, the pendulum swung back harder than ever, like a wrecking ball, clearing space for even more imposing monuments to rationality.

CONCLUSION

Modern advertising developed as part of what James Beniger famously called the "control revolution," crises in industrial capitalism that precipitated movement toward an information society.[88] From the late nineteenth century through the first half of the twentieth century, advertising specialists promoted their scientific credentials and embraced the status of consumption engineers. Agencies and publishers accumulated information to differentiate their services from those of rivals, and leaders of the US advertising industry projected commitments to knowledge making and rationality as badges of professional legitimacy. Especially from the 1910s onward, advertising portrayed itself as a science-in-the-making, a progressive force for economic efficiency. The next two chapters zoom in on an important inflection point in this calculative evolution toward data-driven advertising.

John Bellamy Foster and Robert McChesney locate the emergence of surveillance capitalism in the mid-twentieth century via computerization, militarization, and financialization. The influence of operations research and management science (OR/MS) in advertising sits at the intersection of those axes. The 1950s and 1960s witnessed intensive efforts not only to rationalize executive decision making but also to remodel parts of advertising and marketing around a mathematical science of optimization, formatted for digital computers and adaptive, algorithmic control. Advertisers, agencies, and consultancies absorbed techniques and expertise from the military and the academy, drawing elites from physical and social sciences into corporate America's centers of planning and calculation. Where motivation research attempted to qualitatively probe the interiority of the individual, OR/MS attempted to quantitatively model the whole interlocking world of firms, consumers, markets, and marketing systems. It affected most of all the multilayered processes of media buying and advertisement distribution. These processes grew into power centers of the advertising industry. Through them, the transactions that support media and information systems came to resemble financial markets and automated, high-frequency trading.

The modern paradigm in digital advertising began to acquire institutional and imaginative force through the industry's mid-twentieth-century experiments with OR/MS. Applied to advertising's existing aspirations and challenges, these disciplines shaped and reflected the understanding of what information technology could do, the value of data, the design of markets, and the logistical and knowledge infrastructures supporting a business built around claims, documents, and flows. This little-known episode in the history of advertising paved the way for the forerunners of adtech and surveillance capitalism. It shows how space-age dreams of optimization—still dazzling today—took root in the discourses and mechanics of selling the American people.

3 OPTIMIZATION TAKES COMMAND I: MANAGEMENT TECHNIQUE, FROM THE MILITARY TO MADISON AVENUE

> To manage a business well is to manage its future; and to manage the future is to manage information.
>
> —Marion Harper Jr., president, McCann-Erickson

> Prediction is the goal of all operations research.
>
> —William J. Horvath, Office of the Chief of Naval Operations

Marshall McLuhan began his 1951 book *The Mechanical Bride* with a scathing appraisal of advertising and cultural industries: "Ours is the first age in which many thousands of the best-trained minds have made it a full-time business to get inside the collective public mind. To get inside in order to manipulate, exploit, control is the object now."[1] Over the next two decades, marketing management tried to fine-tune these efforts, enlisting even more of what elite universities, technical institutes, and military-funded science had to offer US capitalism. In 1962 *Advertising Age* welcomed "the new scholarly breed that has joined the advertising fraternity," featuring one of its photogenic members in a flattering portrait.[2] Ithiel de Sola Pool was a professor at MIT and chairman of the Simulmatics Corporation, a company known for its involvement in the 1960 presidential election and, later, for its involvement in the Vietnam War.[3] Simulmatics' "people-machine" used computer simulations of voter behavior to advise the Kennedy campaign about communicating with parts of the electorate.[4] The firm also simulated audience behavior to help advertisers decide how to spend their media budgets.

Advertising Age glamorized Pool's status as a jet-setting consultant—one day doing high-tech demonstrations for major advertisers and their agencies, another day advising the US Navy on classified matters of ordnance. But Pool was just one smiling face in a parade of calculative experts, or "math men,"[5] who were already on the scene when advertising's "mad men" started their creative revolution. The business of influence was absorbing a brave new cohort "composed of mathematicians, scientists and professors in the behavioral sciences who are attempting, with the help of computers, to explore the great unknown areas in advertising."[6] As advertisers and their agencies went on engineering a consumer society, they accumulated mathematical expertise—some of the best-trained researchers in the world—and began to bend their ways of knowing and managing markets and consumers to match the capacities of what some people were calling the "mechanical brain."[7]

Advertisements make up only a fraction of the "texts" marketers produce. Many more texts are created for internal audiences, to document, guide, justify, and narrate business activity. Advertising and marketing are calculative industries, deploying an arsenal of techniques to know the past, predict the future, and manage risk in the breach. Tim Wu calls Arthur Charles Nielsen the "grandfather of today's data geeks," with good reason.[8] As a leading market research firm and an important provider of information to the broadcasting industry, the A. C. Nielsen Company (ACN) was "an extensive and early user of IBM equipment," employing punch-card machines to tabulate data about radio markets since the late 1930s.[9] Eyeing dominance in the audience ratings business, Nielsen was quick to invest in a digital future.

In 1948 ACN contracted with the Eckert-Mauchly Computer Company to purchase "the first commercial adaptation of the Univac Electronic Computer."[10] J. Presper Eckert and John Mauchly were engineers involved in designing the ENIAC, one of the earliest general-purpose digital computers.[11] Eckert, hailed by *Advertising Age* as "the co-inventor of the computer," recognized that the new technology held big implications for marketing, management, and scientific expertise. He speculated that sometime in the foreseeable future, marketing planning would "be handled by a psychologist who is a psychologician of the computer age, fully capable of relating his knowledge of human behavior to the machine"—a prescient if oddly worded forecast of today's behavioral data scientists.[12] The US Army commissioned the ENIAC to automate ballistics calculations. It is irresistible to point out that, since then, advertising's approach to targeting consumers has been characterized by a shift from the indiscriminate blast of a shotgun to the precision of a carefully aimed rifle.[13] Battle terms are common in advertising rhetoric. But advertising's relationship to weapon systems goes beyond metaphor and vocabulary. "As has often been the case," an advertising agency's research director noted, "engines invented for war have later become the tools of commerce."[14]

The first generation of digital adtech emerged at this nexus of defense spending, marketing management, and administrative science. Starting in the 1950s, parts of the advertising business became enchanted with a suite of technologies and technical experts that had been assembled during World War II and continued to enjoy financial backing as strategic assets in the American military-industrial-research complex. Computers, operations research, and related management sciences found a home in consumer capitalism. Together they stood for technologically powered economic and social progress. When writers in the *Harvard Business Review* coined the term "information technology" in 1958, they were referring to the integrated application of computers and operations research—what historian Ronald Kline calls "management technique."[15] That concept passed out of currency, along with inflated hopes for the immediate revolutionary power of new information systems.[16] But during its time at the forefront of US business culture, the definition of information technology as management technique left a lasting impact on the advertising industry.

Operations research (OR) is a mathematical science of optimization that uses models, algorithms, and other statistical means to help executives determine the best possible outcomes when facing complex decisions. It exemplifies the "intellectual technologies" for managing organizations and defining rational action that analysts like Daniel Bell expected to serve as the bedrock for an information society.[17] OR exercised those intellectual technologies within the already knowledge-centric work of selling the American people, and it did so at a moment when various conditions had magnified both the importance of marketing services and the pressures to systematize them. Corporate demand for quantitative methods of decision making helped motivate a lively a subfield of OR dedicated to marketing and advertising. That technical discourse soon inflected particular areas of advertising management, though not without tensions. By 1963, the director of media planning and buying at J. Walter Thompson (JWT) observed that when people talked about the "computer," they were really invoking it as a metonym for the sea change represented by operations research. For him, OR was "a collective term covering all types of mathematical analyses of business problems: in our case, the problem of allocating media dollars in such a way as to maximize the return to the advertiser."[18]

This chapter and the next examine how advertisers and their agencies attacked that problem in the 1950s and 1960s and, in the process, helped establish elements of a calculative culture that prefigured today's data-driven advertising. The advertising business set its sights on optimization and began to imitate operations research and management science (OR/MS). Many of the largest advertisers and agencies in the United States established OR/MS units, and they hired researchers from elite universities as consultants as they implemented mathematical models and marketing information systems. Broadly, their mission was to increase efficiency and rationality in advertising decisions. They tried to achieve that, in part, by eradicating the incalculable— representing the parts of reality they cared about as quantities, probabilities, and formalized expressions. Through mathematical modeling and programming, this emergent class of technical experts formatted reality and decision processes to be amenable to electronic computers and optimal solutions. Advertisers and agencies saw in management technique the possibility to account "objectively" for all the behaviors and processes they wanted to predict, evaluate, and manage. Computerization and OR/MS promised new capabilities to plan rationally, act efficiently, and exploit value and profit opportunities that were inaccessible to traditional procedures. This was, by definition, a project of *quantification* and *rationalization*, and it helped motivate further *datafication* of behavior and social relations, as these math men demanded measurements to fill in or refine their models.

Eradicating the incalculable is a dream, of course. But the tendency to act as if the whole of existence can be counted and computed has had real consequences. Caitlin Zaloom observes a similar process related to the accommodation of electronic technologies in commodity futures markets. Zaloom explains that while the work of designing systems to approximate economic ideals, like rationality, is never perfect or complete, "the long-standing and ever-changing project of creating purely economic spaces is not merely a fantasy. The drive to improve the conditions of economic action leads to the making and remaking of new technologies, reformed rules, creative calculation practices, and emergent classes of professionals who bring the market into being."[19] Advertising's use of management technique was selective, self-serving, and overhyped, but the pursuit of optimization via information technology became a forceful vector in the evolution of media and electronic marketplaces. The calculative evolution did not progress thanks to the undeniable elegance of mathematical proof; it was a political project, where the authoritative language of models and scientific discipline disguised normative choices about what and who counts (and who does the counting).

MATH MEN COMETH

Philip Mirowski calls OR "the unloved orphan of the history of science."[20] Critical studies of media and advertising have scarcely acknowledged its existence.[21] Yet OR/MS influenced theories, design features, practical techniques, and computational dispositions that now power digital advertising. Adtech's mechanics and its ideology of optimization belong to the lineage described in this chapter. Digital advertising platforms are automated incarnations of what operations researchers and management scientists started dreaming up in the 1950s and 1960s.

OR emerged as a disciplinary force in the United States after dedicated refinement of decision-making technologies during World War II. "The war produced a series of 'managerial sciences,'" explains historian Nathan Ensmenger, "including operations research, game theory, and systems analysis— all of which promised a more mathematical and technologically oriented approach to business management."[22] Paul Edwards describes OR as "the first and most successful of the emerging system sciences," adding that it was "largely responsible for the concept of 'optimality' in logistics."[23] While its characteristic techniques were also applied in areas such as agriculture,[24] wartime experiences made an impression on many of the researchers and executives who carried OR's mythology, and its funding, into the cultures of US business and business education.[25]

Computers were a central part of that package, both as calculative tools and as conceptual resources for picturing markets and organization as matters of information processing. "Oddly, a good deal of the early history of AI [artificial intelligence] took place in the basement of a business school," Herbert Simon reflects. That particular business school, at the Carnegie Institute of Technology, was home to operations researchers at the forefront of advertising optimization. "In that setting, and in the decade after 1955, the tools of AI were applied side by side with OR tools to problems of management."[26] Universities and corporations (e.g., General Motors, General Electric, IBM) helped propagate these approaches, as well as an appreciation of their supposed potential, by hosting seminars where executives could learn from operations researchers and management scientists. A brochure advertising one such course at MIT in 1959 described the paradigm shift: "Building on the development of electronic computers, on military advances in understanding decision-making and in using simulation to solve complex systems, and on twenty years of research in feedback control systems, we may now develop new ways to analyze the industrial enterprise."[27]

OR adapted techniques from mathematics and physical sciences to predict and control the behavior of organizations and sociotechnical systems. Its association with respected disciplines provided legitimacy that proponents leveraged to justify its application as a generalizable toolkit for rational planning and decision making. As a systems analysis manager at IBM put it in the *Harvard Business Review*, "It is now clear that the same reasoning processes which have led to notable progress in the physical sciences can be applied to marketing, and that marketing 'laws' can be derived in the same manner as the laws of physics."[28] A 1964 essay about emergent methods for classifying consumers and predicting their behavior suggested that "the probabilistic approach to developing complex marketing models has strong parallels with the methods of nuclear physics."[29] Instead of tracing particles and their collisions, marketing scientists would observe consumers and trace their collisions with all the forces that influence buying behaviors. One marketing professor said the collective methods associated with OR "refer to the attack of modern science on probability type problems which arise in the management, and the control of men, machines, materials, and money in their natural environment."[30] This was a decidedly eclectic pursuit, applying "all branches of scientific endeavor to solve problems." OR groups often "include[d] engineers, logicians, physicists, accountants, biologists, and statisticians as well as various types of business executives."[31] The director of research and information services at a large advertising agency concluded that, for many corporate researchers, "it will be somewhat arbitrary whether they work in the physical sciences or the marketing and advertising sciences since the concepts and tools will appear so similar."[32]

While cautious advocates warned against applying OR to unsuitable problems, many urged an ambitious program of study and intervention. As Edwards explains, "This extension of mathematical formalization into the realm of business and social problems brought with it a newfound sense of power, the hope of a technical control of social processes to equal that achieved in mechanical and electronic systems."[33] As an analyst from the US Office of the Chief of Naval Operations argued in 1948, "this idea—that the scientific method be applied to the study of complex human operations—is as important to the continued economic and social improvement of a technological society as any single physical discovery."[34] Michael Halbert, a marketing consultant and soon-to-be OR specialist at DuPont, defined OR in a way that showcased its capacity to reduce almost any process to a numerical measure of value or efficiency. OR, he said, "considers the system [e.g., a business] as an operating unit, constructs a model (usually mathematical), and attempts to optimize some desired performance characteristic such as net profit, return on investment, machine down time, inventory cost, or the ratio of enemy units destroyed."[35] Simon gave OR/MS a simply staggering remit: "Joining hands with AI, management science and operations research can aspire to tackle every kind of problem-solving and decision-making task the human mind confronts."[36]

By the mid-1950s, OR was "fast becoming" a commonplace in corporate life.[37] "With faith that scientific sophistication amounted to moral progress," one historian explains, business professionals "accepted the claims of [OR's] advocates, believing that science could help them eliminate the use of intuition in decision making, standardize routine decisions, render the future more predictable and controllable, improve planning, and integrate complex operations."[38] OR's industrial migration had powerful supporters and a gee-whiz appeal that excited the press. In 1953 *Business Week* observed "something like a concerted drive by statisticians, professors, and consulting firms to push the use of statistics deeper into business."[39] The statistics described here were not the descriptive tabulations executives had leaned on for decades. These were more exotic inferential techniques for knowing and organizing the world. An accountant at the Rand Corporation took an inventory of OR's repertoire, listing everything but the kitchen sink: "probability theory, symbolic logic, decision theory, queuing theory, linear and dynamic programming, game theory, information theory, Monte Carlo techniques, simulation theory, etc."[40] With these techniques, commercial analysts were "giving hard arithmetical values to things that have always been considered intangible."[41]

OR's presence was first felt in areas of business that closely resembled its military applications: inventory management, production scheduling, financial planning, and logistics, including warehousing and transportation routing. George Kimball explains that "these problems all involve a flow of materials and information through a network of channels."[42] Analysts and executives soon wondered whether advertising and marketing could be defined in similar terms. Firms using OR to rationalize manufacturing and distribution spanned numerous industries, including petrochemicals, automobiles, processed foods, paper products, and more. Some of these companies were among the largest advertisers in the world. For them, there was an obvious appeal to finding new efficiencies in marketing, which was a rapidly growing expense across US and transnational capitalism and an increasingly central element of corporate management.[43]

OR/MS IN ADVERTISING: EARLY APPLICATIONS

Operations research went into business as a "severely practical" discipline of problem solving and decision making.[44] In 1947 Charles Kittel, a physicist at MIT, called OR "a scientific method for providing executive departments with *a quantitative basis for decisions*. Its object is, by the analysis of past operations, to find means of improving the execution of future operations."[45] One of its signature features, he suggested, was that it frames problems in terms of "exchange rates," or "the ratio of output to input for a given type of operation, as measured in suitable units."[46] In other words, it focuses on return on investment. As advocates outlined a program for OR's industrial expansion, they acknowledged the potential to assess and fine-tune the efficiency of advertising.[47] In the first volume of the *Journal of the Operations Research Society of America*, Philip M. Morse of MIT pointed out that "*search theory*, developed for air and sea search for submarines and other vessels, has numerous applications in the problem of the optimum distribution of sales or advertising effort."[48] Kimball, who coauthored with Morse the first US textbook on OR, also saw parallels between military and marketing operations. In the mid-1950s Kimball argued that mathematical models derived from Frederick Lanchester's "formulation of the problem of combat" had been used "with great success to describe the effect of advertising."[49]

Not everyone was convinced of OR's novelty. A researcher at the Leo Burnett agency complained that OR draped "fancy verbal clothing" on what agencies and advertisers already did—apply the scientific method to study markets and consumers.[50] Peter Drucker went so far as to claim that OR's "tools of systematic, logical, and mathematical analysis and synthesis . . . differ very little from the tools used by the medieval symbolic logician, such as St. Bonaventure."[51] But others argued that OR's focus on *systems* and its more sophisticated mathematical techniques constituted something distinct from existing forms of market research.[52] OR was presented as a quantitative and generalizable method for approaching discernibly optimal decisions. Market research was complementary but far more modest and descriptive. Noting operations researchers' tendency to raise questions that market researchers couldn't answer and to agitate for data they didn't have, a professor at the University of Chicago surmised that "if all operations analysts were shot at sunrise tomorrow, company market researchers would live longer and die happier." But the organization would be worse off for the loss, he said.[53]

Some emphasized OR's innovations while also acknowledging its lineage from scientific management. Horace C. Levinson argued this position from an authoritative vantage point. Levinson was chairman of the National Research Council's Committee on Operations Research, which sought to promulgate OR throughout corporate America. Starting in the 1920s, after earning a PhD in mathematical astronomy, Levinson applied his skills to marketing problems. His work for Bamberger's department store in the 1940s was later described by a researcher at the Young & Rubicam agency as "probably the earliest reported attempt at applying operations research to advertising."[54] Without naming Levinson, a pamphlet from the committee he chaired praised the pioneering effort to analyze "the difficult problem of determining the pulling power of certain important types of department store newspaper advertising."[55] Pulling power, here, meant return on investment, or *"the amount of sales produced per dollar of expenditure,"* and the analysis focused on ads that generated an immediate response, so-called R-advertising, which Facebook and others might now call "performance" ads.[56] OR techniques, the pamphlet boasted, brought the problem "to actual numerical solution."[57] The committee made a point of disclosing that these investigations were conducted by a theoretical astronomer, implying that the management of this complex operation owed not just to domain knowledge but also to a scientific attitude and a particular quantitative acumen—a commitment to systematic optimum seeking.

Even earlier, in the 1920s, Levinson had worked for a mail-order retailer. He explained how direct marketing is uniquely suited to optimization via OR:

A small mail-order house offers ideal opportunities for business research or, as it is called today, "operations research." In such a business, all the important data are quantitative. Buying habits of customers are easily studied, for, with few exceptions, every bit of sales action can be traced back to its source, namely, printed matter of some sort that is dispatched to former customers or prospective future customers. Each catalog, each order blank, and each newspaper or magazine advertisement carries its own code designation so that the sales it produces can be readily identified. All in all, the nature of the business is such that it lends itself almost ideally to experimental techniques, so that new and promising ideas can be tested at relatively little cost.[58]

This is a harbinger of digital advertising from one of the first practicing operations researchers in the marketing field. More attachments to direct-mail marketing will surface shortly (and throughout this book).

Publicity credited OR with an impressive record in advertising during the 1950s. Success stories were often drawn from experiences at firms such as Arthur D. Little Inc. (ADL), a venerable research and management consultancy and a vehicle for OR evangelism.[59] In 1949 ADL organized an OR unit to serve industrial clients, in partnership with faculty at MIT and Columbia University, including Morse and Kimball. A decade later, a director from the group reported that they had completed "numerous marketing studies," which generally involved attempts to measure the influence of advertising on sales, statistical analyses of market potential, and studies of consumer behavior.[60] Other consulting firms reported similar experiences. An analyst at the Lybrand, Ross Bros. & Montgomery accountancy said in 1960 that most of his firm's OR work was still of the "bread-and-butter" variety (e.g., production scheduling, shipping and warehousing). "But," he added, "a large number of requests have come in from our clients, and this number of requests is increasing all the time, to study their marketing budgets and the allocation of them. . . . We are assisting some clients to interpret statistical information they continuously obtain, to develop systems of collecting proper information for their purposes and, beyond that, to build models for evaluation of the way they allocate their marketing dollars."[61]

Large advertisers paid particular attention to budgeting. They were intrigued by the potential for mathematical formulas to improve on intuitive judgment or simplistic policies such as setting advertising expenditures as a percentage of the previous year's sales. That sort of budgeting practice was widespread, but it effectively meant that sales caused advertising, and corporations wanted to demonstrate exactly the reverse. Operations researchers claimed their methods could translate evidence of customers' "fundamental behavior pattern" into means of "predicting how the customer will in fact behave."[62] They promised to discern, and thereby manage, the relationships between marketing efforts and marketplace events.

One of ADL's first triumphs supported Levinson's perceptions about the suitability of OR tools for direct-mail marketing and their important continuities with scientific management. OR is about controlling the behavior of systems. Its application to manufacturing and physical distribution tended to be straightforward, since those systems comprised elements that were relatively easy to measure and manipulate—machinery, inventory, and, in some cases, even workers. As companies and their consultants tried to extend managerial control over *marketing* systems, they needed to account for consumer behavior within their calculations. This involved recourse to stochastic and probabilistic models. Operations researchers saw opportunities to analyze advertising effects by conceptualizing purchasing behaviors, such as brand loyalty or switching, as a Markov process—basically, a way of representing the probability that agents in a certain category will make a change in state across some time interval. In this case, that meant representing the probability that customers would stay loyal to the brand they bought last time or switch to a different brand. As analysts from ADL explained, this way of seeing "provides a means for organizing the description of market behavior to permit application of powerful technical apparatus."[63]

In 1950 ADL's OR group was consulting for Sears, Roebuck and Company, whose mail-order catalog was an American icon, sometimes called the "Consumers' Bible." ADL employed operations researchers from MIT, its Cambridge neighbor. In 1956 that group included Ronald A. Howard, a graduate student whose eventual work on Markov decision processes made major contributions to the AI method of reinforcement learning. In 1978 Howard said that ADL's use of Markov processes for Sears was still the only successful application of the technique, and it was this experience that sparked his interest in that area of study.[64]

Sears was trying to determine the optimal policy for mailing its catalogs. The company's existing protocol was basically as follows: Sears reviewed the recent purchase records it kept on file for somewhere around ten million customers, and based on this analysis, the company determined which of those customers to include in the current mailing. The objective was to maximize the expected immediate profit by identifying the most responsive customers. ADL's innovation was to account for profitability over a longer horizon by calculating the probability that sending or not sending a catalog would contribute to a change in a customer's behavioral pattern, such that they would become a more profitable customer. To put it more technically, each customer was classified into one of about fifty categories, based on previous purchases; each of those categories represented a certain amount of profit to the company; and each customer was assigned some probability of transitioning to a different category (and thus becoming more or less profitable) in the future. That transition probability was thought to be sensitive to marketing efforts. The OR solution for Sears was to choose a policy that would increase customers' overall transitions into more profitable categories and thus generate higher returns over subsequent periods.

Importantly, in relation to the prior discussion about the building blocks of a calculative evolution, Howard's recollections stress the importance of Sears' information system. "In fact," he says, "it was the existence of this system that made the entire approach feasible." As Howard describes it, this "unforgettable" system consisted of "two or three acres of green steel filing cabinets" containing customer profiles encoded on punch cards; a workforce of "about a 100 young women [who] continually circulated among the filing cabinets," updating those purchase records, which were attached to Addressograph plates for each customer; and a machine that read the "quantized" information represented by the holes punched in each card and determined automatically whether or not to address a catalog to the customer.[65] The OR group used this sociotechnical assemblage—the machine, Sears' information records, and the labor of its gendered data processors—to calculate the transition probabilities for the Markov analysis and thus decide how to treat each customer based on their predicted profitability.

Other marketers experimented with these sorts of probabilistic methods, although often without the luxury of immediate feedback from identifiable customers. Around the same time, the petrochemical giant DuPont used a similar tack to calculate which consumers or industrial customers could most profitably be targeted with advertising.[66] A research manager from the Benton & Bowles agency viewed this as a productive shift in perception. He told a group of his peers, "perhaps the single most important contribution of Markov chains to marketing is the way in which it forces us to change our point of focus. . . . It's concerned with the *dynamic aspects of market behavior*." He said his agency was "doing a considerable amount of this type of analysis" in 1960.[67]

Formatting consumption behaviors in mathematical notation, management scientists tried to establish relationships between advertising investments and the profits attributable to them, to formalize "optimization calculations" (i.e., choosing the best advertising policy from among several alternatives), and to anticipate the effects of changes in marketing strategy.[68] Early work with Markov processes had definite drawbacks; it relied on unrealistic assumptions, and obtaining data to inform credible probability calculations was not easy or cheap. In fact, it was prohibitive for companies without the formidable infrastructure of a Sears, Roebuck. But these were early efforts to sort and evaluate consumer populations according to their sales potential and the probability that they will act in certain ways. "The Markov process is a simplification of how a consumer actually behaves," MIT's Richard Maffei admitted, "but real data can show how well this kind of process describes what really happens."[69]

Designs to optimize advertising were also passed along in the business curriculum. This book opened with a quote from Albert W. Frey, president of the American Marketing Association in 1961. From 1947 to 1961, the Dartmouth University professor authored three editions of a textbook whose scope is suggested by its title—*Advertising*. By the third edition, its focus fell intensively on advertising and marketing *management*, which meant "attempts to maximize company profits and to meet other objectives by programming the optimum quantitative and qualitative mix of purchase-inducing components in the light of market opportunities."[70] Frey said that executive judgment had been "improved considerably" through recent innovations in high-speed computing, the quantity and precision of available data, and intellectual technologies like "behavioral research techniques, operations research, mathematical programming, systems analysis, and experimentation of various types."[71] Among the most promising applications of OR, he said, was solving optimization problems related to advertising allocation.[72]

ASSEMBLING EXPERT COMMUNITIES

Whatever mathematical method was favored, devising more rational advertising policies was a growing priority for large manufacturers of consumer and industrial goods. *Business Week* reported how DuPont had turned to OR in its search for "a scientific way to fix ad budgets to yield maximum profit."[73] A leader in this effort was Charles K. Ramond, DuPont's manager of advertising research. Ramond went on to serve as the technical director of the Advertising Research Foundation (ARF), a professional association created in 1936 by the Association of National Advertisers and the American Association of Advertising Agencies. Ramond was also the founding editor of the ARF's *Journal of Advertising Research*, established to reinforce advertising's scientific credentials and address disciplinary concerns. Writing in the journal's first issue in 1960, Ramond took stock of the mysteries confounding advertisers (following the convention of presuming that anyone in authority in business is a man): "Who knows the return on his advertising investment? His optimum media combinations? His right themes?" With these animating questions, Ramond framed the journal's agenda in terms that appealed to the mode of knowledge pursued at DuPont. "The task of the marketing scientist," he wrote, "is the prediction of consumer behavior." Through experimentation and refinement, management could gradually solve any lingering mysteries and irrationalities: "there will simply be better advertising decisions—measurably better."[74]

The ARF was a catalyst for this work, hosting regular liaisons between applied and academic researchers. Its fourth annual conference in 1958 focused on mathematical techniques for understanding the economic impacts of advertising. For some speakers, this was a chance to hand out congratulations. John F. Magee, a senior analyst at ADL, crowed about his firm's successes: "we have worked out ways of predicting the effect of advertising changes on profits and these ways are being used. Ten years from now, I believe we will pretty generally be able to predict the effect of advertising on profits and we will be able to use these predictions to design efficient advertising programs."[75] For other speakers, however, this meeting was an opportunity for a wake-up call.

Jay W. Forrester, a professor of industrial management at MIT, told the attendees that they needed to do better. He complained that "much of so-called advertising research is itself merely advertising."[76] Noting that advertising "is a powerful and important influence in our present-day economy," he stressed the geopolitical urgency of scientific progress: "The challenge and new frontier in our capitalist society during the next three decades is not space flight, but the science of management and economics. It is in management and economics, not on the moon or Mars, that the current international competition will be won. The American corporation is the heart of the American economic system. How well we fare will depend on how well American corporate management understands its job."[77] Forrester saw advertising as critically important but in desperate need of rationalization. "I doubt that there is any other function in industry," he surmised, "where management bases so much expenditure on such scanty knowledge."[78] Forrester insisted that "intuition is totally unreliable" for the management of dynamic marketing systems. "We must turn instead to methods developed in the study of engineering and military-weapons systems."[79]

Forrester spoke as an admitted outsider to the advertising industry. Others at the conference had more skin in the game. Russell L. Ackoff and John D. C. Little represented an OR group at what was then the Case Institute of Technology. They were already consulting for large marketers such as M&M Candies.[80] Ackoff and Little promised no panaceas, but they expressed "the firm conviction that many of the basic problems of advertising strategy can have more than a little light thrown on them by the type of application of scientific methods that has come to be known as *Operations Research*."[81] They used these methods to measure advertising effects and return on investment and to formulate procedures for selecting the most efficient media placements. Companies were intrigued by OR's potential, even if they did not grasp all its details. An executive from Standard Oil of Indiana reportedly said of Ackoff, "I'm not sure what he's selling, but I want to buy some of it."[82]

Throughout the 1950s, then, a distinctive professional community began to establish some particular means for determining advertising effectiveness and optimizing media selection. More precisely, two communities that were committed to solving these respective problems—advertisers and advertising agencies—invested in technical experts and knowledge-making practices that promised authoritative solutions. A class of management scientists who applied their methods to marketing was taking shape in research firms and universities and hiring out their skills to industry. Before the end of the decade, they found additional institutional support and some organized opportunities for exchange.

THE ARF'S OR DISCUSSION GROUP

From 1959 to 1965 the ARF convened an OR discussion group in New York City. Leading US companies were represented up and down its membership.

The group's first chairman, Michael Halbert, was an OR specialist at DuPont and had previously been a research associate at the Case Institute of Technology. Across a dozen meetings, the group assembled delegates from firms spanning much of the American economy (table 3.1), as well as staff from the Department of Agriculture and the Weapons System Evaluation Group of the US government.

Table 3.1
Partial list of organizations represented in the ARF's OR discussion group.

Consumer goods	General Mills Johnson & Johnson M&M Candies National Biscuit Company Pillsbury Scott Paper Company
Advertising agencies	BBDO Benton & Bowles D'Arcy J. Walter Thompson Ogilvy, Benson & Mather Young & Rubicam
Accounting/management consultancies	Alderson & Associates Arthur Andersen and Company Arthur D. Little Inc. Lybrand, Ross Bros., & Montgomery McKinsey & Company Price Waterhouse
Computer, research, information, communication/media	CEIR General Electric IBM Rand Corporation RCA Laboratories Simulmatics Corporation Time Inc.
Petroleum/petrochemicals, pharmaceuticals	Atlas Chemical Industries Imperial Oil E. I. du Pont de Nemours & Company (DuPont) Smith, Kline & French Socony-Mobil Oil Union Carbide Corporation
Higher education	Columbia University Harvard University Massachusetts Institute of Technology Princeton University Stanford University University of Pennsylvania

Participants discussed a range of OR projects conducted by their organizations, although corporate secrecy inhibited an open scientific exchange. More generally, the gatherings promoted the use of "mathematics as an aid to management decisions."[83] As of 1961, more than one thousand transcripts of the group's first five meetings had been requested by researchers and managers, which the ARF considered "remarkably high interest in matters so complex."[84] By its ninth meeting in November 1963, the group had forty-three members.[85]

Information, science, and technology were once again part of the repertoire advertising could draw on to assert professional authority in US capitalism—at a time when the efficient management of consumption seemed like an urgent necessity and a compelling reason to leverage or invest in technoscience. By the middle of the twentieth century, as one observer reports, it was well established that "accurate knowledge of the characteristics of consumers and markets, of the relative effectiveness of appeals and media, is indispensable to present-day advertising and marketing."[86] Advertisers demanded research services, and almost all agencies claimed to be providing cutting-edge expertise. Agencies' research activities varied in rigor and verve, but it was common sense by then that companies needed "aids more scientific than one man's opinion" to cope with the scale and complexities of marketing and media businesses.[87] Because clients essentially shared decision-making responsibilities with their agencies, including the ultimate determination of how to spend clients' money, those agencies also absorbed pressures toward rationalization and optimum seeking. "Research personnel need thorough backgrounds in statistics and the behavioral sciences," a 1961 textbook advised.[88] An article in the *Journal of Marketing* later confirmed that "research departments in advertising agencies have taken the lead in developing algorithms that produce optimal solutions to media mix problems."[89]

CALCULATIVE AGENCIES

Computerization and OR shifted the composition and culture of the advertising business. An appetite for data and analytical muscle and an increasingly scientific cadence in management values were reflected in the development of research services, which were growing at a faster rate than advertising services overall. From 1950 to 1964 the amount agencies spent on research, as a proportion of total expenses, increased by 50 percent.[90] The largest agencies led this general trend, as well as the particular gravitation toward management sciences (table 3.2).

Table 3.2
Partial list of top advertising agencies, 1955–1970

Agency	Year formed	Domestic billings (millions) / Rank				Notable clients
		1955	1960	1965	1970	
J. Walter Thompson	1877	$172 / 1	$250 / 1	$351.5 / 1	$436 / 1	• RCA • Scott Paper • Lever Bros.
Young & Rubicam	1923	$166 / 2	$212 / 4	$306.1 / 2	$356.4 / 2	• Johnson & Johnson • General Foods • General Electric • Procter & Gamble
Batten, Barton, Durstine, & Osborn (BBDO)	1928	$162.5 / 3	$232 / 2	$292.7 / 3	$324.4 / 3	• DuPont • American Tobacco • U.S. Steel • Republican National Committee
McCann-Erickson	1930	$132 / 4	$216 / 3	$270 / 4	$246.5 / 7	• Colgate-Palmolive • General Motors • National Biscuit Company
Leo Burnett Company	1935	$69.2 / 5	$116 / 6	$184.7 / 6	$283 / 4	• Kellogg's • Marlboro • Pillsbury
Benton & Bowles	1929	$68 / 6	$114 / 7	$130 / 10	$139.3 / 15	• Procter & Gamble • Philip Morris

Source: Advertising Age.

Young & Rubicam (Y&R) formed a dedicated OR unit in 1961. Before that, the agency was engaged in what its research director Peter Langhoff called "the seat of the pants operations research school."[91] By the mid-1950s, Y&R began experimenting with OR methods for media selection. Martin K. Starr, a professor at Columbia University and a consultant to Y&R, reported on some of these experiences in a textbook coauthored with David W. Miller, a colleague at Columbia and later a consultant to JWT.[92] Starr had been invited to Y&R by Langhoff, who reportedly admitted, "I don't know what OR is all about, but I'm sure that it's something we need."[93] Langhoff hoped OR would increase the presence of the physical sciences in advertising research, and Y&R brought Starr into the office on a weekly basis starting in around 1956.[94] Starr helped develop a computer-supported media selection algorithm, and he collaborated for decades with Bill Moran, a research manager at Y&R who had previously worked at Pillsbury and later became the director of market research at Lever Brothers—two giant advertisers and early advocates of OR/MS.

BBDO invested in OR around the same time Y&R did. The agency hired Charles L. Wilson as its director of research in 1959, and Wilson hired David B. Learner a year later. Learner began developing a linear programming (LP) approach to optimal media selection, working with computing and data resources from CEIR. LP is an optimization technique for finding the maximum or minimum value of some target criterion, or objective function, subject to constraints. Learner had been exposed to LP in his previous job at General Motors, one of many companies that hosted seminars with marketing and business researchers. One of the seminars Learner attended was led by Abraham Charnes of Northwestern University and William W. Cooper of the Carnegie Institute of Technology. Charnes and Cooper went on to rank among the most decorated management scientists in the United States. Charnes was short-listed for the Nobel Prize in economics in 1975, and Cooper was the founding president of the Institute of Management Science (which later merged with the Operations Research Society of America). In 1982 they shared the John von Neumann Theory Prize for their contributions to OR/MS. Learner met them again in 1961 at a summer course at the University of Michigan, and soon thereafter the group went to work on an LP-based media optimization model.[95] Charnes and Cooper collaborated with BBDO researchers for many years, funded in part by the Office of Naval Research, the US Army Research Office, and the National Science

Foundation.[96] They helped BBDO refine later iterations of its media optimization model (LP II) and devised predictive models for marketing new products. By the late 1960s, they proposed that they work on retainer as continuing consultants.[97] Minutes from an April 1968 meeting show that the agency's first priority for Charnes and Cooper's ongoing efforts was "immediate adoption of LP II as a working component of day-to-day operations in BBDO."[98]

BBDO established a Marketing Science Department in 1971. It was founded by the agency's director of research services, an MIT-trained Bayesian statistician who eventually became the CEO of BBDO Europe. He described the department in retrospect as "a remarkable business commitment to the utility of OR in Marketing."[99] A senior vice president of marketing sciences later claimed that BBDO was "participating in the expansion of behavioral statistics," building directly on work at major centers such as the University of Michigan and Bell Labs.[100]

As mentioned earlier, the J. Walter Thompson Company saw OR and computers as interlinked. In an internal report dated January 1, 1963, management discussed the computer system the agency had ordered and the OR techniques it might use to take advantage of this new technology. JWT expected to use the machines for media planning and consumer research, as well as for routine accounting and administration. The agency had already retained the consulting services of Professor David Miller, whose writing partner, Martin Starr, helped Y&R develop its computer-based optimization model. However, the company expressed misgivings about what its peer agencies were doing: "We have not been satisfied that any of the much touted approaches to the use of electronic data equipment in media planning would accomplish what was claimed and, in fact, might dangerously cause reliance to be placed on mechanical findings from assumptions fed into the machines which are actually not subject to quantitative measurement."[101] Despite that sober note, by the late 1960s JWT had implemented an "integrated system for media planning," which culminated in the "computer production of an optimized schedule"—pretty closely resembling the techniques its rivals used for media selection.[102]

A 1970 memorandum from JWT's New York office acknowledged the need for both "a functional unit of specialists in psychology, statistics, operations research, etc." and specialized quantitative help from technology and information service vendors.[103] By then, JWT's London office had already organized an Operational Research Unit ("operational research" was the

British term). Its director explained in 1967 that "the method and the discipline" of OR could be applied to many marketing operations "with at least equal effectiveness" as in its uses for production, distribution, and inventory control.[104] He expected OR to increase the intensity and sophistication of market research and analysis. "Operational Research," he said, "feeds on data and already a massive amount of information has been collected about markets, and marketing operations which will serve as a base on which to work. . . . [I]ncreased use of operational research will lead to calls for more market data and probably for more precise and detailed knowledge of markets than has been available in the past." The culmination of applying OR to marketing would be the construction of a "total marketing model" that could specify relationships and predict outcomes with unprecedented detail.[105] His vision was consistent with active discussions about using management information systems and computerized "nerve centers" for adaptive control.[106]

This is an important reminder that OR/MS work was not exclusive to the United States. British operational researchers designed some of the first quantitative advertising models, although they focused almost exclusively on print advertising, reflecting international differences in media industries. For example, researchers from London's International Wool Secretariat developed an early model for "optimum geographical distribution of publicity."[107] Drawing inspiration from the natural sciences, other researchers tried to model advertising effects based on analogies with physiological stimulus-response and epidemiology.[108] With the help of a physicist from the Royal Naval College, they developed an "electronic simulation of advertising response" to predict coupon returns and assist with media planning.[109]

Without diminishing the international scene, this book focuses on US approaches to advertising optimization, which were the most widely publicized at the time. Even a researcher from the London Press Exchange admitted in 1970 that BBDO's linear programming model was the "best known [mathematical] formula approach" to media selection and that Y&R's offering appeared to be "the most ambitious of all media models."[110] (Both are discussed in the next chapter.)

By the mid-1960s, then, OR/MS techniques were becoming ordinary and authoritative in parts of the advertising industry. "Over the past decade," the director of information systems at Benton & Bowles wrote in 1966, "the roles of Management Science and Operations Research have grown, under the salient influence of computer methodology, from a modest experimental project state to a full scale systems state." Researchers and managers, he noted, were tackling larger and more complex problems with new analytical resources: "Techniques such as mathematical programming, simulation, Bayesian analysis, multivariate analysis, and Markov analysis have become everyday production and research tools to the Advertising Industry."[111]

Some saw broad significance in what agencies were doing. Writing in *Journalism Quarterly*, the manager of research development at Leo Burnett celebrated the "role of advertising research as a catalyst" for systematic knowledge and techniques of persuasion.[112] He admitted that advertising "seems like a most unlikely spawning ground for a new behavioral science," noting that "a good deal of advertising research practice remains today what it has been in the past—a pseudoscientific rationalization for advertising expenditures." But he detected a transformation: "The successes of modern cost accounting, production planning, inventory controlling procedures and other 'management science' methods have created a new climate in American business. Businessmen feel a new need for rational understanding of the business processes for which they make major expenditures." With all this disciplinary expertise being assembled around practical business problems, "the polyglot jargon of behavioral science and communication theory is being distilled to produce a pragmatic language for decision makers."[113]

POLYGLOT JARGON: A LANGUAGE OF POWER

Management technique venerated certain classes of problem solving and problem solvers. "With the computer comes a small army of mathematicians and scientists, new terminology, new media outlooks and attitudes," one journalist observed. "As with anything new and unknown, there also comes fear."[114] As Y&R's Langhoff put it, "today's prophets are heard voicing the new vocabulary—the vocabulary of management science. For marketing men, this verbal veneer can have a disturbing, even a threatening ring."[115] One such prophet, a senior vice president at the Kenyon & Eckhardt agency, looked forward to the emergence of "a 'marketing elite' . . . primarily concerned with the new social, behavioral and mathematical sciences," who would be fluent in "sophisticated techniques" such as "operations research models."[116] A 1960 profile remarked with some fascination on his habit of using words like "'feedback,' 'maximize,' 'input,' 'output' and 'optimum.'"[117] When *Sponsor* took stock of "the new vogue in media vernacular" a few years later, it listed "automation," "data processing," "linear program-ming," and "memory drums" among the terms of the trade.[118]

Seasoned professionals sometimes resented this jargon's invasive influ-ence. Puzzled and annoyed, the director of TV and radio at a Detroit agency pulled together a partial lexicon of this "almost mysterious language." He called it "a form of cant" used by specialists in media, research, and other areas of the business "to make what is said or written unintelligible to per-sons outside the group." Among many examples, he noted the tendency for media buyers and sellers to take inspiration "from the language of war-fare."[119] Similarly, an executive at General Mills disparaged the growing trend toward "almost automatic" media selection, calling it "Malarkian Media." He was skeptical of specialists in "numerical manipulation" who were rising through agency media departments. "We all know that, in spite of mountains of numbers to the contrary, media is nowhere near the exact science it is fre-quently made out to be." Shielded by the dazzling promises of the computer age, "Malarkian Media blots out questions of substance with little columns of numbers."[120]

Critics of management technique were understandably defensive. Ratio-nalization threatened the autonomy and status of advertising professionals who benefited from their personal relationships with customers and clients, the perception that they possessed unique instincts and experiences, and the use of intangible metrics that left enough room to dodge responsibility for disappointing or unprofitable campaigns. A mathematical science of optimization implied that these subjective factors could be objectified and commanded by management systems. The naysayers quoted above recog-nized that this new language and numeracy were levers of power, confer-ring a cultural legitimacy on these specialists and protecting their claims from critical adjudication by nonexperts. Computer analysis, for example, implied an intimidating gravity. According to the media manager at Y&R, "There seems to be something very final and very impressive about the computer's output. The little printed numbers often appear to overrule the soundest judgments of the most capable marketing man, even when the data that produced the numbers are open to serious question."[121] One observer pointed out that some managers without mathematical expertise tended to "treat operations researchers with a 'halo effect.' Individuals who can manipulate symbols, use computers, refer to the proper mystic and mathematical concepts, and arrive at solutions are accorded status."[122]

These anxieties indicate the struggles surrounding change. Agency media and research departments were experiencing a local version of broader shifts in quantitative knowledge making.[123] Some regarded the prominence of mathematical models and the imitation of "more mature disciplines" as indices of marketing's elevation into a legitimate "branch of empirical science."[124] Philip Kotler, for example, welcomed the language of "higher mathematics." But he also recognized "the danger . . . that mathematics might be used by some to lend authority to some essentially ill-conceived decisions." He warned against careless or cunning uses of jargon: "Marketing men will be subject to further mathematical name-dropping in written reports and at their conventions—cybernetics, information theory, econometrics, distributed lags, Bayesian decision theory, and so forth. Although these terms stand for perfectly good ideas, they should be viewed as part of a larger plan to advance knowledge, and not just represent verbal glibness."[125] A market researcher at Xerox agreed, writing in 1966, "'OR' and 'mathematical models' have become catchwords; they have become 'in' types of activities. In our world of market research, where the norm is to use semi-scientific techniques in inappropriate ways, people are particularly vulnerable to the latest gimmick that will lead them from the wilderness of their own lack of self-confidence."[126] Some hoped that as corporate executives were "increasingly drawn from the ranks of science and engineering," advertising agencies would no longer be able to "abuse" statistics and self-serving research.[127] But as a former management scientist from BBDO put it, "We all know how to bludgeon managers into quiescence with technotalk."[128] The present state of affairs in digital marketing suggests that "technotalk" remains a forceful bludgeon.

Even sympathetic observers complained that OR was overhyped and misused. The ARF's Charles Ramond was a fierce proponent of using science to bolster the credibility of advertising. He even recommended the imitation of physical sciences as a means of "virtue-by-association." But he also admitted that advertisers usually lacked vital knowledge about the efficacy of their investments. This "chronic information deficiency" left them exposed to any discipline promising solutions to their problems, an affliction he called "galloping panacea." He called OR the latest and gravest "epidemic" of this sort. Its apostles seemed to be peddling "a magic formula" for advertising, heralding the eventual replacement of "flesh-and-blood practitioners with the smug certainty of decision-making computers."[129]

This was a reasonable critique. OR presented itself as a mode of technocratic optimum seeking that could be generalized to solve almost any decision problem. Not surprisingly, that attitude created conflicts within organizational hierarchies. "What we are seeing," explained a market research director at Pillsbury in 1962, "is the development of the 'elite of logic and mathematical precision' because of the nature of mathematical and computer technology." While he celebrated the rising importance of these elite mathematical experts, he feared that their new technologies would not be exploited to the fullest unless OR was adapted to the practical realities of management.[130] Marketing managers often lacked the time, wherewithal, or even inclination to search for truly optimal solutions. They would usually just *satisfice*, or settle for a less-than-perfect choice that met some minimum threshold.[131] One study concluded that a decision maker is typically content with "a solution he can justify (or rationalize) to his superiors, 'sell' to his peers, and pass on to his subordinates with the assurance that they will consider it logical and acceptable enough to act upon it in a predictable way."[132] Operations researchers faced the reality that acceptance of mathematical designs would be won not simply by their logical merits but rather on the basis of organizational politics, culture, and trust. According to the market researcher from Pillsbury, "The important thing is not the rightness or wrongness of the recommendation, but how the decision-maker feels about it."[133]

For people like Donald Longman, director of research at JWT and American Marketing Association president in 1963, the treatment of OR as a "fad" was impairing its development.[134] The patience and rigor required to build better models, collect better data, and refine OR in advertising competed against the day-to-day demands for problem solving from companies and managers. "The best thing we can do for our customers," one consultant said, "is the kind of research that will lead to better decisions later, rather than putting out fires today." But he acknowledged the counterpressure arising in commercial technoscience: "If your boss's boss comes in and says, 'I want to know what's wrong in Topeka with the ad campaign we started last week,' you can't very will [*sic*] say, 'Don't bother me, I'm doing a ten year research project.' You try to find out what happened in Topeka."[135] Likewise, DuPont's Halbert strongly endorsed efforts to gather better information to feed into computer models, but he conceded, "we can't sit back and wait for the millennium when we have perfect data."[136]

Other critics were more devastating. A. S. C. Ehrenberg referred to OR modeling and optimization methods as "Sonking, or The Scientification of Non-Knowledge."[137] He lampooned OR with a series of "scientific laws," including, *"The Law of the Ignorant Problem-Solver,"* described as someone with no understanding of the topic but an irresistible urge to optimize, and *"The Law of Perpetually Promising Pseudo-Probabilistic Paraphernalia,"* where analysts fish for insights with impressive but inappropriate statistical methods.[138] Referring to stochastic models of purchasing described by several management scientists, Ehrenberg wrote, "if these authors actually knew anything factual about consumer behaviour, they have successfully kept it both from their readers and from their models."[139]

These barbs may have raised eyebrows at the Royal Statistical Society, but it is not hard to imagine an industrial consultant shrugging them off. "The operations research worker," wrote one such consultant, "typically is willing to use any or every analytical technique at his disposal which shows promise of helping solve his problem."[140] OR consultants were offering their clients what Bernhard Rieder calls "interested readings of empirical reality."[141] Careful theory building and even causal explanation sometimes conflicted with that ruthlessly practical mission. One marketing professor admitted that picking predictive techniques based on convenience, rather than any motivating theory, "is not to be belittled in solving a company problem where the criterion of goodness is, 'Does it work?'"[142]

This orientation prefigures some hallmarks of commercial data science, where the search for useful patterns and correlations motivates indiscriminate data collection and promiscuous analyses. Understanding why something happens often takes a backseat to predicting what is likely to happen. Recall from the start of this chapter that "prediction is the goal of all operations research." That attitude followed OR into the advertising world. As ADL's Marcello Vidale explained in 1959, "I think that market models are predictive and do not necessarily imply causal relationships. A good model is one that makes correct predictions."[143] Kenneth Longman, an advertising researcher, agreed: "it's generally true, that what we are looking for are relationships in nature of a statistical or correlational type; relationships between how much you expend on advertising and what you get in return in sales."[144] Not quite a decade later, Longman, then an editor at *Management Science*, suggested that marketing was less like a theoretically driven science and more like a form of engineering that "has the task of using whatever applicable theory it can find, coupling it with some hard trial and error, and developing a method for solving a problem."[145] As one professor put it to the ARF's OR discussion group, "I don't think we ever know whether a model is right. We only know whether it is useful."[146]

Looking back from the mid-1980s, a group of prominent operations researchers that included Charnes, Cooper, and Learner argued that scientific progress had come not from "substantive theorizing" but from new methods of collecting and analyzing marketing data. They likened this to the invention of microscopes, providing access to relationships and phenomena that were previously hidden. Combining new technologies of data collection with multivariate statistics, OR/MS was developing "new and improved ways of dealing with large scale data systems" and "better types of decision support systems." The next step was to move beyond "pure prediction" and toward intervention and control: "management's role is to change the way the future is likely to behave relevant to a set of goals or standards. . . . Marketing strives for maintenance and growth of profit making in a largely uncontrollable environment, using whatever controllable factors managers have available."[147] Advertising technoscience carried forward the dream of using technology to transform more of reality into controllable factors. OR/MS thereby helped establish marketing optimization as a design value for a capitalist information society and its large-scale data and decision systems. When an anonymous marketing data scientist described their work in 2018, they echoed exactly the rhetoric that accompanied OR/MS: "I'm doing my job right if I can give my company and stakeholders the data they need to make decisions. . . . You're solving a puzzle every day. How can I predict future behavior? . . . Throughout the history of mankind, people have loved fortune tellers."[148]

CONCLUSION

In a 1961 presentation A. C. Nielsen Jr. relayed some of the questions he was hearing from executives: "How can my ever-increasing marketing costs be lowered in relation to results?" "How can marketing results be more accurately determined and predicted?"[149] Nielsen was a top tycoon in the marketplace feedback business—someone that captains of industry looked to for informed answers. And he had a hopeful story to tell them. "While much progress is yet to come in the area of reducing marketing to a science, we can point to the very real progress that has been made to date." His specific advice for improving profitability sounds familiar today. He encouraged executives to build their marketing programs around a strong "intelligence system" to continually monitor consumer demand and an ongoing, "carefully administered program of controlled experimentation"—A/B testing different marketing plans and levels of expenditure, iterating toward an "optimum."[150] Sixty years later, he might have told them just to advertise on Facebook. He also expected that existing "intelligence systems for reporting the activities of consumers" would soon be "augmented by other systems for finding out what the consumer thinks, or hopefully, finding out what he might think tomorrow."[151] Today, that sounds like the self-promotions of Google and Amazon. In 1961 it was basically a nod to computerization and OR/MS—to information technology as management technique.

Math men sold the advertising business on visions of efficiency and rationality. The overarching impulse was to optimize advertising by determining the relationships between advertising actions and marketplace outcomes. This was a holy grail for advertisers, and it was what brought OR into advertising management at corporations like DuPont. But it was an uncomfortable project for their agencies. There was a serious risk that close examination of advertising effectiveness would lead to the conclusion that advertisers should spend less. Since agencies received almost all their revenue from commissions on clients' media spending, they had no interest in discovering that an advertiser's budget was too large. Agencies therefore had to spin the power of computers and OR in a different direction. They concentrated on performing more efficient ways to spend advertisers' money. As advertisers' appointed decision makers, agencies used management technique to promise an unprecedented scientific capacity to choose.

According to a textbook by consultants to BBDO and the Leo Burnett agency, "The problem of selecting from a nearly infinite number of possible combinations was solved to a considerable extent when John Mauchly and J. Presper Eckert submitted a plan to the U.S. Ordnance Department in 1943 for an electronic computer." The authors noted that high-speed computers, and the optimization methods they enabled, were perfectly suited to one of agencies' major functions: planning and buying media placements.[152] Operations researchers working for advertising agencies framed the media selection problem as a type of decision making that lent itself to mathematical formalisms and optimal solutions. This framing relied heavily on models, computers, and a rising technical elite. It affected language, dispositions, data requirements, and organizational relations in advertising and media businesses. Media directors and researchers at advertising agencies saw the science of optimization as a powerful set of resources that allowed them to make authoritative claims that would validate their expertise. More broadly, advertising agencies saw new tools and discourses to build their media services for the future that corporate America wanted: computerized, rationalized, optimized. They channeled management technique toward the efficient buying of consumers.

4 OPTIMIZATION TAKES COMMAND II: THE RULE OF CALCULATION

On November 16, 1961, BBDO bought a full-page advertisement in the *New York Times* to announce a breakthrough innovation in media selection. The agency promised to find the best possible advertising placements to satisfy clients' objectives by using a method called linear programming (LP). LP is a mathematical approach for determining how to maximize or minimize the value of some criterion, such as profit or cost. BBDO was using it to create media schedules that would maximize advertising exposure among specified audience segments, given budgetary and other constraints. "This new technique, which utilizes high-speed computers," the ad explained, "amounts to giving media men a power shovel to work with instead of a spade"—a claim BBDO's president repeated later that day when the company unveiled this marvel of operations research (OR) at a meeting of the American Association of Advertising Agencies. According to BBDO, its use of LP was such a dramatic advance in industrial art that the agency had a duty to publicize the method. "We make this offer because we believe that advertising is a serious business upon which much of the strength and prosperity of the American economy depends."[1]

Charles Wilson, the director of research at BBDO, hailed this innovation as a departure from the advertising industry's "spotty history" of incorporating scientific advances into professional practice. "For every pound of real OR study that's actually been conducted," he complained, "there has been a ton of speeches and papers by people outside the advertising industry asserting that it ought to use OR more effectively." BBDO aimed to close that gap by seeing like an algorithm, "transforming the everyday operating processes and procedures of media selection from plain language into the language of mathematics, then into equations and groups of equations." Configured this way, the realities and decision problems facing the agency's media staff could be acted on with computers at superhuman speed and scale. For all that splendor, the payoff was a straightforward appeal to efficiency: "The general purpose of this mathematical programming approach is to allocate limited resources in such a way that a business can obtain maximum profit. The media purpose is to allocate a given budget for maximum advertising. In short, the purpose is to get the most advertising for the dollar."[2]

It would be hard to overstate the fanfare surrounding BBDO's demonstration. Looking back on that day, the media director at J. Walter Thompson wrote in 1967, "The fallout from the 'nuclear' explosion that followed has affected the cellular structure of every media department in the business."[3] BBDO's ad declared the end to "a media rule of thumb," a phrase that plays on multiple meanings of "rule"—as a measure, a governing principle, and a reign of power. BBDO was promising a new regime: a rational order rooted in science, technology, and speed. The *rule of convention* would be succeeded by the *rule of calculation*.

Management science (MS), and particularly optimization techniques like mathematical programming, appeared perfectly suited to the needs of agencies' media departments. These ways of knowing and ordering the world could help media directors both with the services they were selling— systematic decisions for wringing the most value out of clients' ad budgets— and with the challenges they were facing—more media, more choices, and more demands for efficiency. The fit seemed obvious. As operations researchers at MIT put it, "The problem of preparing a media schedule is a natural one for the application of models and computers."[4] OR/MS professionals approached the media selection problem as one of optimum seeking. "Any method of media selection is concerned with deriving an optimum or near optimum schedule," explained an analyst from Arthur D. Little. That meant finding "a feasible schedule that maximizes the numerical value of some objective function."[5] Media planners had always worked intuitively to meet marketing objectives. What operations researchers did was to formalize the "objective function"—the criterion for evaluating a media schedule—as an explicit, quantitative statement. They configured media selection as an algorithmic science of decision making. When a consultant and former mathematics instructor at the US Naval Academy took stock of successful business applications of OR in 1963, he put "selection of advertising media and expenditures" at the top of the list.[6] A decade later, business professor David Aaker described media selection models as "probably the oldest, the most developed, and most readily accepted of all the management science models."[7]

This chapter looks at how agencies used management technique to optimize media planning and buying—the operations involved in deciding how to spend advertisers' money and distribute their ads. OR/MS framed this area of the advertising business in ways that reflected corporate demands for efficiency and rationality and that favored styles of technical authority that were closely related to computerized data processing and emergent information systems. A burgeoning "mathematical elite" used optimization models to organize advertising decisions for rational discrimination, allowing managers, planners, buyers, and analysts to quickly and "objectively" discern differences in the value of audience segments and the media used to reach them. This meant that the processes for evaluating variables and alternative choices, and for reaching value-maximizing decisions, had to be expressed formally and quantitatively. It was, by design, a project of rationalization, and it implied strong pressures toward datafication of the behaviors and relationships that advertising professionals wanted to include in their models. Perhaps above all, optimization helped those professionals perform expertise. The rule of calculation gave decision makers a persuasive repertoire for justifying costly choices and taking credit when objectivity yielded new or surprising efficiencies. In both its technical and symbolic dimensions, this iteration of adtech contributed to an understanding of media buying and advertising distribution as operations that should be managed with the help of automated, data-driven decision systems.

MODEL MEDIA: WHAT OR/MS SEES WHEN IT LOOKS AT MEANS OF COMMUNICATION

As historian Ronald Kline explains, "Mathematical modeling was a hallmark of the postwar 'behavioral revolution' in the social sciences."[8] This movement is evident in advertising and marketing research. From the mid-1950s onward, journals, textbooks, and conference proceedings all teemed with mathematical models built to represent advertising decisions, marketing processes, consumer behavior, information flows, and much more. Academic enthusiasm did not translate automatically into adoption by corporate managers, though, so industry groups took steps to evangelize OR/MS techniques. Around 1960 the Association of National Advertisers commissioned Robert S. Weinberg to prepare a report detailing the use of mathematical models for advertising planning. Weinberg was IBM's director of market research and a veteran optimum seeker. He had been an economist at the Department of Defense and a member of the US Air Force's Project SCOOP (Scientific Computation of Optimum Programs)—an initiative that applied computers to "management, budgeting and logistics problem[s]" led by George Dantzig, who famously devised the simplex algorithm for linear programming.[9] Weinberg later served as chairman of the Advertising Research Foundation's OR discussion group. After leaving IBM in 1966, he became the vice president of corporate planning at Anheuser-Busch. A 1971 profile described him as "probably the highest positioned management scientist in American industry."[10]

Weinberg's work on advertising models started from the premise that competitive pressures and the magnitude of advertising investments were pushing corporate management "to become extremely selective in choosing among various marketing alternatives."[11] Models could provide clearly structured rules and metrics for making those discriminating decisions. "Our purpose in building mathematical models," he explained, "is to describe the way in which the marketing mechanism operates. If we know the quantitative characteristics of this mechanism . . . we shall be able to project, with a specified level of probability, the inputs, in terms of dollars, physical quantities, man-hours, etc., required to maintain a stipulated level or marketing activity or to attain a given objective."[12] Weinberg was, in this instance, instructing *advertisers* how to use models to plan expenditures, rather showing *agencies* how to allocate clients' budgets, but this work indicates the appetite for rational management among large advertisers and the pressure on agencies to keep pace. A review of Weinberg's report called it "mathematical missionary work."[13] *Television* magazine stated that it "demonstrates what many experts agree might be the way of marketing and media planning in the not-so-far-off future: the use of mathematical models, or operations research techniques, as a means of relating advertising costs and market share to profits."[14]

A SIMPLE ANATOMY

Generally speaking, analysts and decision makers use mathematical models to represent the components and relationships defining any process, system, or part of reality they want to predict and manage. The elements accounted for within the models are expressed numerically, as quantities, probabilities, coefficients, and so on. Basically, these models are mathematical statements of purpose and priority—formal renderings of how (some part of) the world works, what things and relationships are considered important, what their value is to the decision maker, and what measurements are required to control the behavior of the system.

In the context of marketing, mathematical models allow managers to carefully define the decisions they or others in their organization must make and to assess the available alternatives in commensurable units of value. Having specified their understanding of the situation, the logic of how it works, and the values they want to maximize, executives can rationally, and often algorithmically, determine a course of action. The outcome of solving the equation(s) making up the model is basically a prescription for how to manipulate one or more variables under management's control so as to achieve the highest payoff (or lowest cost). These are intellectual technologies for organizing information, and the realities that information represents, in a calculable form. And they can be used to discipline knowledge work because, at least in principle, management can count on subordinates to make consistent choices when their decisions are structured by the same model.

Even in the mid-twentieth century there were many types of models. An optimization model of the sort used in advertising has some typical components. Most importantly, it has an "objective function"—a unit of value that the modeler and manager want to maximize or minimize. It is an "objective" (or "criterion") because it represents some goal or performance measure; it is a "function" because that unit of value is presumed to vary in relation to other variables in the model. To give a simplified illustration: If we assume that sales volume depends on the amount of advertising investment, then we could say that sales are a function of ad spending. And if we knew the nature of that functional relationship (i.e., how a change in spending corresponds to a change in sales), then we could determine the level of advertising investment that would maximize sales.

The media selection models examined here are generally more modest, in that the objective function represents some measure of *advertising exposure*.

This is a convenient criterion because it is presumed to correlate with the sales and profit outcomes advertisers really care about, while being easier to measure and manage. Generally, a model's decision problem is formulated as one or more differential equations, and its decision variables are the initially unknown values that are populated in solving the equation(s). In the case of media selection, decision variables are usually the quantities of advertising units purchased in each media option. The idea is to find the values for these decision variables that optimize the objective function. Solving the model shows how much should be spent on each media option to maximize exposure with a defined audience.

But decision variables are subject to limitations on their possible values. Constraints include the advertiser's budget, the supply of advertising inventory, contractual obligations, subjective preferences (e.g., a client's desire to advertise, or not, in a particular medium), and logical guardrails that prevent a model from making absurd recommendations, such as telling a media buyer to purchase a *negative* quantity of inefficient ad placements. The input data for the models include ostensibly empirical measures, including the size and composition of the audience for a given media unit and its price, and estimates and parameters derived from professional judgments about how effectively each medium will carry an advertiser's message and influence consumers.

PURPOSE AND PRIORITY

An interesting implication of these designs is that they forced model builders to explicate their values and objectives. We therefore get a glimpse at what marketers and OR/MS experts saw when they looked at advertising and media. In a 1958 paper about predicting advertising results, an operations researcher explains the point of the whole enterprise: "The results that businessmen are interested in are profits. I know there are many legitimate reasons why business supports advertising programs, but in most cases the result that is looked for is improved profit through an improved sales position."[15] This view was not universal in OR/MS, but it was undeniably predominant. As two MIT researchers explain in their textbook *Management Science in Marketing*, "advertising should be viewed as an investment, and the goal of advertising should be to maximize the returns on investment."[16]

This disposition leads to an understanding of media as extensions of the marketing system and as means for the efficient organization of consumer populations as investment opportunities. OR professors and ad agency consultants David Miller and Martin Starr define media in their 1960 book as "the instruments by means of which communication or advertising strategies can be fulfilled."[17] They go on to elaborate that the *audience* is the key element: "a medium is nothing more than a communication channel that can be observed. One of the main distinguishing features, therefore, is the audience with which each channel communicates." Miller and Starr also identify a "great variety of demographic and psychological factors," such as income and intelligence, that can be used to classify audiences: "Occupation, years of schooling, number of children, social poise, degree of tension, credit standing, and frugality are just a few examples of other possibilities."[18] And they urge specificity in defining which audience to target:

> The decision-maker selects certain characteristics which he believes are important attributes of potential customers. He must also be prepared to specify how much of each characteristic is desirable. It is not enough for him to state that intelligence is a significant variable with respect to the sales potential of the consumer; he must also specify that some intelligence level—say four, rated on a scale from one to ten—characterizes the best level of consumer intelligence for his purposes. In doing this, the decision-maker is choosing what he believes to be the most favorable state of nature.[19]

A British operational researcher who worked for several large advertisers proceeded from a similar orientation. Advertising campaigns, he explained in 1962, typically aim to either "persuade some portion of the population to take some definite course of action, or to create and maintain an idea or an attitude in their minds." This is obvious enough, but then he stresses the importance of media distribution: "Whatever the objective may be, the essence of advertising is that a message must be conveyed from the advertiser to some class of people whom we call the target population."[20] Another operational researcher noted that computing machines process the world through a similar optic. Describing the potential for media planners to combine data from different media and devise all-inclusive media plans, he explained that the computer "regards a media vehicle as simply a probability of delivering a certain weighted impression to an individual, and . . . indifferently weights individuals by demographic or purchasing or attitude characteristics."[21] As the director of print media at Young & Rubicam put it, computerization enabled commensurable measures of "apples and oranges," as even different media could be "compared in terms of their vitamin content."[22]

In a classic critique of audience measurement, Ien Ang shows how audience ratings produced "a discursive framework" that enabled the television industry to "know its relationship to the audience" in a statistical idiom.[23] Optimization also produced a discursive framework. Experts classified and materialized audience segments and the media inventory representing units of their attention as quantitative values and probabilities—as information that could be handled with computers and mathematical formalisms. Optimization models forced analysts or managers to explicate their objectives, strategies, and values and to clearly identify the data needed to calculate rational choices. A close look at OR/MS in advertising shows that this technical community regarded media as useful mechanisms of social sorting, enabling the efficient distribution of ads to apparently profitable populations.

OR/MS thus brought particular ways of seeing and counting to the processes of evaluating consumers and circulating advertising money throughout media systems. Mathematical modeling formalized rational, algorithmic procedures for ranking and discriminating among a sometimes staggering number of alternatives. Working through all the possible investment choices available to advertisers within a limited time required "a systematic technique for calculation."[24] This was exactly what agencies promised to provide with their media services. They defined media planning and selection as an applied science of decision making.

DECISION MAKING AND DECISION MAKERS

Commercial applications of OR/MS were intended to improve executive decisions. As marketing theorist Wroe Alderson put it, "an executive with a problem is an individual responsible for decision and beset with uncertainty as to the course of action to follow. To solve a problem is to reduce uncertainty to a point that will permit a choice to be made."[25] Marion Harper Jr., president of the McCann-Erickson agency, considered the "virtual explosion of research . . . in technology, operations, and marketing" around the start of the 1960s as an "investment in management decision-making—to help determine the future environment for a particular course of action, or to indicate the superiority of one course of action over another."[26] By 1965, the research director at Young & Rubicam noticed "an increasing demand from top management, particularly during the past decade, for more rational decisions in the marketing area. . . . The tolerable range for error has shrunk to disconcertingly narrow limits. Consequently, means of reducing uncertainty are urgently sought."[27]

Management technique promised a salve for this anxiety. OR/MS imported into advertising a particular repertoire for defining problems and their solutions—what sociologists might call a "technological frame."[28] In particular, operations researchers positioned media selection as terrain for decision theory, which Kenneth Arrow defined as "a formalization of the problems involved in making optimal choices."[29] OR/MS framed agencies' media planning and buying activities as mathematical and engineering problems that lent themselves to modeling and optimum seeking, an ostensibly progressive contrast to traditional styles of intuition and negotiation, such as the fabled three-martini lunch. The rationalization of media selection—as procedures to optimize a formally defined objective function—made new demands on the organizational and calculative environment. It called for (and reacted to) technoscientific expertise, more abundant and detailed behavioral data, and expanding facilities for rapid computation and communication. Managers were trying to make this kind of advertising work legible for algorithmic, machine-supported decision making. Under the rule of calculation, media selection was configured as processes that could be expressed mathematically and executed programmatically.

Compared to other advertising operations, the allocation of budgets seemed like a great candidate for formal treatment. Media selection was information intensive and punctuated by frequent decision-making events. Those decisions involved high levels of choice and uncertainty, yet they had to be handled as routine matters. In a 1955 textbook, a lecturer at Columbia University writes, "One of the most important services the agency offers its clients is the selection of media best calculated to carry the advertiser's message. This service entails considerable knowledge of individual media and the markets they represent."[30] Calling the media buyer, or time buyer, "a key controller of advertising dollars," the vice president in charge of media at Leo Burnett said, "Ever since broadcasting came into prominence as an advertising medium . . . the principal job of the timebuyer has been decision making."[31] BBDO's vice president of research agreed: "Buying time or space is a decision making process."[32]

The media function evolved in tandem with data processing and analysis. Agency media departments were already calculative spaces, buzzing with information and the industrial media for managing it. Describing the experience of navigating BBDO's media department, the director of advertising at Pepsi said, "I'm quickly lost among the calculators, ARB's, Nielsen's, SRDS's, salesmen, availabilities, worksheets, overtime slips and all the other elements that go into one of the busiest departments in the agency."[33] The industry registered "major tremors" in the early 1960s, as media departments appeared to be "entering the space age."[34] Much of the excitement centered around computerization and OR/MS techniques, which had been "developed at a cost of hundreds of millions of dollars in the weapons and space race," as a systems scientist who had worked at MIT's Radiation Laboratory noted in *Television* magazine.[35] Enthusiasts welcomed "a new age of scientific and efficient media planning," while skeptics "dismissed the computers as a publicity gimmick." In any case, "the dialogue on computers" led to contests over the status of information and technical expertise.[36]

The potential ascribed to computers and scientific innovations elevated the professionals associated with them. The media managers and research directors who worked on OR projects were cast as apostles of a progressive, rational future and paraded through the spaces and publications where the industry assembled to talk about itself. As one journalist put it in 1963, "The media man with a computer is taking on new stature in the agency field as his calendar fills out with speaking engagements to address his fellows and the media multitudes on the wonders of a new age."[37] Media specialists staked claims to new prestige, expertise, and responsibilities, or in some cases, they had those thrust on them by other executives who were eager to capitalize on the cultural hype around futuristic technoscience. "It's generally agreed," *Sponsor* reported, "that with an ever increasing volume of progressively more sophisticated data on its way to aid media departments in performing their functions more effectively, more sophisticated analysts and researchers will be needed, and that segment of the department will take on greater stature."[38] As one executive forecast in 1964, "Tomorrow's media buyer will operate . . . as strategist and tactician. His new importance will result from improved and expanded media forms and the accumulation of instantaneous automated data on media and markets."[39] Computer systems programmed to execute mathematical analyses would rapidly convert data into recommendations about how to adjust advertising strategies and optimize media plans. "In that highly automated era the logical candidate to assume authority for use and control of this complex is today's marketing oriented timebuyer—the media specifier with his unique insight into the almighty power of the ad dollar."[40]

Using management technique to climb the status hierarchy required some finesse. Not everyone agreed that media selection could be as formal and precise as a computerized science of optimization seemed to imply. Some media personnel were actually empowered by treating it as an art. This preserved the intangible value of their unique experience and judgment, protecting them from automation or the encroachment of efficiency experts. They could deflect accusations of irrationality by insisting that their work defied any strict scientific accounting. The media director at Grey Advertising wrote in 1960, "as yet, we simply do not have a scientific body of knowledge about media selection." Importantly, though, he stopped short of denouncing scientific aspirations. "All business is an art. Media selection, in fact, with the data now at its disposal, probably comes closer to being a science than any other brand of the advertising arts."[41]

Despite these tensions, industry discourses were clearly normalizing the perceptions that decision making in advertising required a scientific disposition and that media departments were poised to take advantage of powerful new means of obtaining and using information. Media selection was becoming increasingly complex. Planners and buyers had to choose from more and more publications, broadcast stations, and combinations of advertising opportunities. Experts liked to illustrate the computational burden with colorful vignettes. As the manager of Y&R's media department put it, "a media buyer selecting 10 media from a group of 100, has 17,310,000,000,000 different alternatives available to him." Analyzing one alternative per second around the clock, it would take half a million years to evaluate all the choices.[42] IBM's Weinberg gave an even more dramatic demonstration. By permitting additional decision factors, such as discretion in the magnitude of spending (rather than just choosing whether to use a certain medium), he portrayed the analytical challenge as truly astronomical. "If we had 10 different media alternatives, and we wanted to decide how much to spend on each, and we were willing to allow one of a hundred possible expenditure levels, and wanted to literally search each combination, we would have to look at a hundred quintillion possibilities." If each possibility were encoded on an IBM punch card, he said, "we could construct 113,140 piles from the earth to the sun."[43] Commonsense constraints would rule out the vast majority of these options, but the universe of plausible choices could still number well into the millions.[44]

Helping decision makers cope with these complexities was part of the pitch that operations researchers peddled on Madison Avenue and that agencies repeated to their clients. They also promised to unlock hidden value through more massive and rational data analysis. Proponents claimed that these models and computer methods could produce media plans that defied traditional norms and intuitions but were quantifiably more efficient at satisfying marketing objectives. The following sections examine some specific designs.

MEDIA OPTIMIZATION MODELS

LINEAR PROGRAMMING AT BBDO

BBDO worked with prominent operations researchers and the data processing firm CEIR to develop a computerized media optimization model. It was based on the "Dantzig simplex" algorithm for solving a linear program.[45] The basic idea behind staging media selection as an LP problem, explained consultants Abraham Charnes and William W. Cooper, was to spend the advertiser's money on a schedule of media placements that would "achieve the maximum total impact on a target audience."[46] More formally, the objective function in BBDO's optimization model—the criterion it was designed to maximize—was called "rated exposure value."

"Exposure" meant that potential consumers saw or heard an advertisement. What made those exposures "rated" was that the media planner (1) assigned a weight or score to each audience segment, representing its sales potential and its relevance to the advertiser's marketing plan; and (2) assigned "impact values" to the available media options, based on qualities that could affect consumers' responses to advertising (e.g., strength of a broadcast signal, editorial tone of a magazine, need to demonstrate a product in color or with moving images). Based on the worth of audience segments and the expected effectiveness of media vehicles, as well as audience ratings indicating the distribution of those segments across those media, each advertising option could be scored in terms of rated exposure value. Assembling a schedule that maximized rated exposure value within the allotted budget then became a straightforward optimization problem, and "the simplex algorithm provides an exact and unique solution."[47] The algorithm quickly searched the available alternatives to find *one best schedule* that satisfied whatever constraints were specified (e.g., spending a certain part of the budget on television or reaching a certain percentage of people in a specified income class).

BBDO's David Learner claimed the computerized LP model generated schedules that delivered 30 percent more rated exposure value than media planners were able to achieve manually.[48] The calculations themselves were not intractable to manual computation, but electronic data processing dramatically accelerated their pace and, together with the LP technique, increased by orders of magnitude the permutations of media placements , that could be evaluated within the time available for planning and buying. *Business Week* reported that this system let BBDO "quickly solve problems of advertising media selection that formerly took weeks or even years."[49]

In addition to recommending the optimal schedule, the LP output included a "sensitivity analysis" that indicated to what extent the result was influenced by the constraints, parameters, and estimates entered into the model. With sensitivity analysis, the LP system not only told media personnel how to optimize a schedule within predefined constraints but also indicated how changes in those constraints, as well as other variables, could affect the objective function. These calculations quantified the extent to which corporate advertising policies, which might be motivated by tradition or instinct, were inefficient when measured against the stated objective. Sensitivity analysis also revealed how precisely certain values needed to be specified, and thus whether it was necessary and economical to collect more data.

In a sense, LP provided an accounting of a decision, testifying to its rationality or documenting where it deviated into irrationality. If an advertiser insisted on placing ads in a prestigious publication because it flattered top executives, or if a company followed the more mundane but still arbitrary policy of investing a certain percentage of its budget in magazines, the LP's sensitivity analysis could specify the cost of those choices. Learner observed that this information "is really at the heart of the management decision-making value of the model."[50]

Although LP sounded impressive, it was no magic formula. "Because the problem of media allocation can be so readily formulated to conform to the problem structure required by the linear programming algorithm," wrote one researcher, "this approach was among the first tried by advertising agencies."[51] Its application in advertising was later described as "a classic case of a technique . . . looking for a problem."[52] A business professor at Columbia University complained that LP was "often misused as a publicity vehicle for ad agencies."[53] Another observer worried that both advertising and OR would suffer "if [LP] models are used so that an agency can say, 'look how scientific we are.'"[54] Even admiring analysts acknowledged that because BBDO so "aggressively publicized" its model, rivals and skeptics were inclined to decry the entire effort.[55]

Despite LP's provenance and BBDO's claims about its performance, there were many doubts about how well it worked. Critics pointed out that media selections guided by LP often differed little from intuitive judgment, and its output depended on assumptions and estimates made by the advertising professionals who evaluated audience segments and media vehicles.[56] Indeed, it was possible for media planners to set the parameters such that the model's output would justify whatever they wanted to do. The head of research at Ogilvy, Mather & Benson said, "We tend to think that when we punch data into a machine, it becomes fact. Actually a lot of the information put in here is purely subjective."[57] According to a systems analyst from IBM, "Analytical optimization techniques, such as linear programing, frequently employ unrealistic simplifications as with a linear objective function (which is often not linear) that maximizes media exposure (which often is not the central problem)."[58] As he and many others noted, it seemed implausible that there was a linear relationship between the volume of advertising exposure and the value or effectiveness of advertising (i.e., that each exposure was equal to the ones before and after it). This assumption, one researcher remarked, "is like saying that three girls aged 20 are equivalent to a woman of 60."[59] BBDO denied that its use of the method assumed strict linearity, but the perception of this deficiency stuck. Operations researchers tried using nonlinear and dynamic programming to better account for the realities of media selection, but even large computers were still overwhelmed by the requirements for solving these problems by the end of the 1960s.[60]

Despite these controversies, technical experts portrayed mathematical optimization as an important *cultural* advance. Herbert Maneloveg, BBDO's media director, explained that the purpose of using LP was not just to calculate but also to achieve a more calculative disposition. The aim, he said, "is to reorganize and discipline our thinking, to make sure our precious few advertising dollars go where they can do the most good and show our clients exactly where it goes."[61] Mathematical modeling and optimization were part of a rationalizing and quantifying project. According to an internal BBDO memorandum, LP "helps (the media planner) by making him get a clear picture of his objectives and his own judgmental evaluations."[62] "Because we are dealing with mathematics," Learner explained, "everything that we can or cannot do, each kind of data, each management decision must be precisely stated."[63] BBDO was codifying media decisions, a crucial precursor to automating those decisions.

New York University professor Darrell B. Lucas also highlighted the broad significance of mathematical modeling, despite its limitations. Lucas helped BBDO develop the LP system, and he defended its method of accounting numerically for estimates, unknowns, and *qualitative* factors: "What is the value of so magnificent a mathematical formulation if it still has so much in it that is subjective? . . . In the first place, it forces us to systematize our thinking, and to quantify those vague values which have been both our weakness and our protection in the past. Secondly, the use of this procedure enables us to make the utmost gain from such facts as we already have; and we do have some substantial facts."[64] The state of the art would progress, Lucas argued, as researchers marshaled these facts and methods "in a continuous attack on areas where once we had only subjective judgements as guides to values."[65] DuPont's OR specialist, Michael Halbert, likewise praised the effort, noting that "the explication of subjective techniques" was itself "a significant breakthrough."[66]

BBDO's use of mathematical programming helped make media selection legible and executable by computing machines. It also revealed limitations in the way media planning problems had been defined, and it exposed the inadequacy of available data for populating these equations. It made particular demands of knowledge and analysis that fit with the mathematical thrust of management technique. Professionals at other agencies shared these perceptions about the reformatting of their decision space. "Because the computer is a logic machine," JWT's media director explained, "it has forced us to define our audience objectives more carefully, to proceed in a rational step-by-step fashion throughout the analytical process, and in the end has made us face up to some hard decisions in evaluating media which we used sometimes to evade."[67] The director of information systems at Benton & Bowles also observed that even imperfect applications of optimization methods like LP catalyzed shifts in data creation: "Advertising and Marketing Research directors began to call the Management Scientist in at the early stages of the design of their experiments. . . . As a result, *some* data collected began to take on the right form for the model; and *some* models began to predict with *some* measure of reliability. This in turn encouraged *some* more attempts at model building—so a cyclical evolution had begun."[68]

YOUNG & RUBICAM'S HIGH-ASSAY MODEL

Researchers at Young & Rubicam experimented with linear programming in the late 1950s, but they were unwilling to tolerate the simplifying assumptions needed to compute a mathematically "best" schedule. In consultation with Martin Starr of Columbia University, Y&R developed what it called the high-assay media model. It was an algorithm—a set of sequential instructions—that told planners how to spend an advertising budget. The firm likened it to gold mining: "Think of the mines as being different media, and the gold as sales prospects."[69] The model was designed to mine all the available gold from the medium that best met the advertiser's goal and then move on to the next best choice.

High assay belonged to a class of "iteration models." It assembled a schedule in a sequence of steps, each time selecting the media option that returned the most value per dollar—or "the lowest cost per prospect obtained," as Kenneth Longman put it.[70] Making the optimal choice at each discrete decision, high assay was an example of a "greedy algorithm."[71] This was technically a nonoptimizing, heuristic approach, since it did not necessarily arrive at the best choice for the entire decision set. Rather, it aggregated a bunch of locally optimal choices, achieving *near* optimization with much lower computational demands than LP. But Y&R did not always bother to make this distinction. It told *Advertising Age* that the computer would analyze data with a programmed "decision system" and then produce a media schedule with "optimum reach, frequency, and periodicity."[72] Joseph St. George, a manager of media and computer systems at Y&R, said, "If our decision system is correct, if we have completely accurate media data and ad impact data, and if we have made an exact appraisal of our prospect and market situation, then we should have confidence that our computer will produce the perfect recommendation."[73]

The system's objective was "minimizing the average cost per customer obtained."[74] To do this, media planners input three types of data into the agency's computer: (1) consumer-product behavior (e.g., brand share, prospects' demographics, probability of switching brands, customers' purchase rates); (2) consumer-media behavior (e.g., audience size and composition, rate structure for each medium, duplication of audience members across media options); and (3) consumer-advertising behavior (e.g., effect of ads on brand switching or loyalty, effect of added exposure over time). The name "high assay," as one observer later reflected, conveyed the notion that some audiences are worth more than others, "that a marketer's top priority is to find ways to identify and discriminatingly reach these high value target customers."[75] According to St. George, this was accounted for by relating data about media audiences to "market-prospect data which contains information about who we wish to sell, where they live, how easily they are persuaded, etc." Like BBDO, Y&R recognized that the state of the art left much room for improvement, but the agency saw this approach as putting its media department on a more rigorous footing. "Realistic appraisal tells us that at this stage we do not have all the information and data we need," St. George admitted. "But, used judiciously, these models can help us pinpoint areas in which we must make subjective judgments, and can force us to be much more scientific about ways in which we invest our clients' ad dollars." He concluded that "decision models are going to be a fact in all our lives in the years to come."[76]

J. WALTER THOMPSON FOLLOWS SUIT

J. Walter Thompson entered the OR field with less publicity than its rivals, but it quietly kept pace. When JWT acquired its RCA computer in 1963, the agency also acquired Norman E. Sondak, a data processing manager from RCA, to oversee the installation, and he stayed on to manage JWT's computer operations. Like the other experts discussed earlier, Sondak believed computer usage demanded exacting, quantitative thinking. "If one can't express a problem in numbers, one really knows very little about it," he said in the company newsletter.[77] Sondak and the computer arrived not long after JWT hired an OR consultant to optimize media selection.

By the late 1960s, JWT had organized an "integrated system for media planning."[78] According to a written summary of the system, the first step in the agency's planning process involved using multivariate analysis to divide the consumer population into market segments, "each with a specific purchase probability."[79] Identifying the inter-correlations of variables that most powerfully predicted certain buying behaviors (e.g., age, gender, income, education), this "automatic interaction detector" method produced a tree-branch arrangement of increasingly narrow subgroups, many of which had to be reaggregated to be practical targets for media buyers. "With this information," the document explains, "we can get a very accurate picture of the relative value of various kinds of people."[80] Staff then used one or both of JWT's computerized selection models to pick a media plan that would efficiently satisfy some specified objective. One model evaluated alternative schedules that were already assembled; the other automatically chose the optimal combination of media options.[81]

The agency continued to build computer-based optimizing tools in subsequent years.[82] According to a bibliography circulated to media personnel at JWT's New York office, from 1970 to the end of 1973, the agency's Media Analysis Department produced eleven reports or manuals pertaining to the computer resources used for media functions. These covered online computer systems, data processing procedures and services, and various media-specific programs.[83] The department boasted in a 1970 memo that "the informational and technical resources available to our media people are unmatched in the advertising business."[84]

SIMULMATICS' SIMULATION

As full-service agencies sharpened their calculative tools, specialized vendors marketed rival products and services. Perhaps none piqued more interest than the Simulmatics Corporation. In the early 1960s Simulmatics used a computer simulation method to evaluate media plans for large marketers such as Colgate-Purina, General Foods, and DuPont.[85] The company fabricated a hypothetical sample in the form of a database representing the US population and its media habits. The sample included 2,944 synthetic individuals whose probability of being exposed to an advertisement in a given media vehicle was defined as a function of their socioeconomic and demographic features. Using "an electronic brain that performs 250,000 separate calculations per second," Simulmatics' "media-mix" model simulated those individuals' media behaviors to forecast the audience exposure of a given media plan.[86] The company claimed this enabled advertisers to realize media spending efficiencies of up to 30 percent.[87]

Computer simulation did not optimize media selection. It was a tool for evaluating schedules that media planners had already assembled. An operations researcher from RCA contrasted it to LP by describing simulation as a technique for finding a solution when "the algorithm for improving the system is missing."[88] Instead, the media-mix program predicted the likelihood of reaching desired consumer targets with alternative plans. The computer reportedly output as many as four thousand tables of information detailing the media consumption of Simulmatics' data-based population.[89] The media-mix model also classified consumers in ways that went beyond basic demographics. "For example," Simulmatics' Alex Bernstein explained, "for each individual, what kind of car does he own and how old is it? How does he or she distribute purchases between supermarkets and drug stores? Between grocery and drug stores? Does he live in a hard or a soft water area?"[90]

James S. Coleman, a professor at Johns Hopkins University and a member of Simulmatics' research board, expected computer simulation to precipitate a "revolution in the development of advertising and marketing techniques."[91] The company even planned to "shoot for the ultimate goal in advertising prediction: forecasting a campaign's sales effectiveness."[92] This chutzpah was characteristic. Chairman Ithiel de Sola Pool boasted that Simulmatics "pioneered in the application of computer techniques to the market place. After a great deal of intricate research and the accumulation of tremendous amounts of data, we devised a mathematical method to evaluate the probable reactions of consumers to the introduction of a new product or advertising campaign."[93]

Simulmatics made a big splash. But Pool's pioneering claim was a bit flattering. The company was not alone in recognizing the potential of computer simulation. To give one colorful example, William Fair, cofounder of the Fair Isaac credit scoring company, described an elaborate plan for programming a computer to run simulations of business decisions to identify the most profitable courses of action. "Essentially," he explained in 1953, "the proposed computer is an apparatus for making an analogue model of the economic world centered on the business in question. The entrepreneur describes his world (quantitatively), chooses his policies, and turns on the machine. The machine then runs 'into the future' at the high speed permitted by modern design, and delivers the answer on what will happen if the model fits reality."[94] Consultants at Arthur D. Little reported using computer simulation as early as 1958 to test "alternative advertising campaigns on a hypothetical customer group."[95] Around the same time, Richard Maffei at MIT used simulation to help Heinz with distribution logistics, and he claimed to be working on similar methods to study the relationship between the company's advertising and its profits.[96]

Advertising agencies—together with computer service vendors like CEIR—were also entering this field, albeit after Simulmatics. As Jill Lepore points out, "these agencies had better data than Simulmatics" from their own market research units, some of which had been established decades earlier.[97] A few agencies even used simulation methods to fill gaps in their databases. Y&R trademarked something called the Data Breeder Model, which was used to make estimates "where existing data are inadequate." Starting with known information about consumer behavior, the computer "produce[d] the best estimates of missing information."[98] Just like Simulmatics, Y&R used this system to predict how a media plan would reach specified populations.[99]

Clearly, the advertising industry saw promise in what Simulmatics was doing. To some observers, simulation represented a productive integration of behavioral and management sciences. Weinberg gave Simulmatics some wry (and prescient) praise: "One of the things wrong with advertising research is that the psychologists got hold of it. This may be dwarfed by the damage the mathematicians are going to do—I am not sure—but it seems to me that somewhere along the line, someone has to marry the mathematicians to the psychologists." Simulmatics' media-mix model, he said, "provide[s] a very interesting device for doing this."[100] Others saw simulation as valuable for understanding consumers and audiences. We tend to think of media in this period as exclusively organizing people into undifferentiated "masses," but media managers appreciated that computers and new methods of analysis could help them operationalize finer and more flexible consumer targets. As a vice president of media at Benton & Bowles explained in 1962, "Simulation opens a new horizon and places emphasis upon individuals rather than households."[101] That same year, the president of CEIR claimed that analyzing a "statistical equivalent of the entire consuming public" would allow advertisers to move closer to their true objective of maximizing profits: "with the introduction of new analytical techniques and the new computer machinery, achievement of the delicate balance necessary to optimize profits may become a reality."[102] Computer simulation was a charismatic medium onto which the advertising business could project its dreams for the future.

MEDIAC: A MODEL UTILITY

The final media decision system examined here was born on the same campus as Simulmatics. At MIT in the 1960s, John D. C. Little and Leonard Lodish developed a number of marketing and media selection models. Little joined the School of Industrial Management in 1962, and he is thought to be the first student to earn a PhD in operations research.[103] Funded by the US military and large corporations such as Mobil Oil, Little's research helped extend OR/ MS into a subfield called marketing science. Much of his work focused on computer-based adaptive control systems, "real black boxes for a technical management of markets," as Franck Cochoy puts it.[104] Essentially, these systems assembled and analyzed information about marketing environments, organizing the basis of managerial decisions. Lodish was Little's student and had previously been an OR analyst for the Mead Corporation. He joined the Wharton School at the University of Pennsylvania in the late 1960s.

Little and Lodish designed models for optimizing (or nearly optimizing) media selection via dynamic and heuristic programming. They worked under the auspices of Project MAC, a large research and development initiative, well known for its contributions to computer time-sharing, sponsored by the Department of Defense's Advanced Research Projects Agency. Little and Lodish characterized their main innovation as "a media planning calculus," which was embodied in a program called MEDIAC (Media Evaluation Using Dynamic and Interactive Applications of Computers) and accessed using a terminal connected to a time-shared mainframe. "By a 'calculus,'" they explained, "we mean a system of numerical procedures for transforming data and judgments into a schedule. The model supplies the structure, the user supplies the data and judgments, and the computer supplies the muscle."[105] Perhaps because of their affiliation with Project MAC, Little and Lodish hailed MEDIAC as a testament to the power of computers, which they called "enthusiastic clerks," capable of examining many more advertising alternatives than was possible with manual calculations.[106] They compared MEDIAC favorably to existing media selection models by noting several unique features: it accounted for advertising effectiveness "in greater detail"; its "on-line" computer system afforded continuous querying of a database via the remote console's user-friendly interface (which meant not only more timely information but also the possibility that media personnel could learn to understand the model through frequent human-machine interactions); and it was computationally efficient and thus economical in terms of the cost of computer time.[107]

The model's objective was to maximize total market response to advertising—a criterion representing the sales expected to result from the level of advertising exposure achieved by a certain media schedule. To start, the population was divided into customer segments and characterized according to the "sales potential" of the people in each group. Each segment was also defined by its media habits, allowing planners to estimate the distribution of market segments across media options. Like other operations researchers, Little and Lodish held that "the main purpose of media is to deliver messages to potential customers efficiently."[108] Therefore, they evaluated media based on their "exposure probabilities" among the people in various segments and their "exposure value." Exposure value was a subjective quantification; it represented the presumed power of an ad placed in a given medium to increase the disposition of someone in a given market segment to buy the product.[109] And, unlike some other models, MEDIAC accounted for the diminishing impact of ad exposures over time and at high frequencies.

Calculating each media unit's "increment of response per dollar," MEDIAC helped media planners assemble a schedule that maximized sales potential across market segments. Selection could be optimized with dynamic programming if planners were considering just one or two market segments; beyond that, computation time became prohibitive, so a heuristic search was used to find a near-optimum solution.[110]

Little and Lodish described their goal as increasing advertising "productivity." They claimed to improve media planning by 5 to 25 percent, as measured by the model's definition of market response.[111] "Most media planning is rather macroscopic," they said, and focused mainly on "simple efficiency measures" such as cost per thousand exposures. "We want to show how more phenomena can be handled with greater ease than these usually are today."[112] Computer time-sharing provided the capacity to account for more "facts and phenomena" within the media calculus. In its first few months of operation, MEDIAC was used to allocate "several million dollars of advertising."[113]

In addition to their research, Little and Lodish are notable for their entrepreneurialism. They founded a company, Management Decision Systems, and Little claims that he recruited clients for their consulting work from the summer seminars MIT hosted to teach executives about OR/MS.[114] They sold the MEDIAC model to a firm called Telmar, a leading vendor of computerized media planning and analysis tools—what might be considered an early adtech company. Founded in 1968 by a media supervisor and director of data systems at Y&R, Telmar was "staffed by a group of bright young computer wizards," according to the *New York Times*.[115]

Observers saw the commercialization of MEDIAC as a catalyst for the wider adoption of OR/MS techniques. Two other management scientists at MIT described MEDIAC as a type of "model utility," an already assembled program or algorithm that customers accessed through a data communications network, or "computer utility." They expected these complementary utilities to enable many more firms to "usefully and economically take advantage of the computer and marketing models."[116] At least some working managers agreed. By 1970, marketing research and information system specialists at Coca-Cola reported that "the capabilities of the computer now allow us to use many statistical techniques which have been around for some time but which have not really been feasible due to the large amount of computation required." Explaining that "statistical and operations research analysis methods are available in the form of several software packages," they specifically praised Little and Lodish's company and its products.[117] At least thirteen agencies reported using MEDIAC for media planning in 1982, although only a couple were still using it in 1994.[118]

A textbook from the early 1980s described MEDIAC as an important advancement in media planning models, both because it accommodated more intricate details and because it was packaged as a complete decision model available for "on-line access via a remote terminal." Unfortunately, its sophistication probably limited its adoption: "it may be too complex for most media planners."[119]

TOWARD A RULE OF CALCULATION

These models were ahead of their time. It would take a broader set of enabling conditions to furnish the data and commercial relations necessary to really exploit these techniques. But early experiments with OR/MS were agents of that change. They represent a loosely organized attempt to reformat the infrastructures and political economy of knowledge in advertising and to link corporate aspirations to dreams of mathematical optimization. They prefigure the culture of data-driven advertising and its orientation toward surveillance, profiling, and discrimination. Essentially, optimum-seeking approaches to media selection tried to identify value more precisely; express that value more formally and in commensurable, quantitative units; and thereby facilitate rational, algorithmic choices. Management technique looked at audiences and advertising opportunities in ways designed to help planners discern new actionable efficiencies. Practically, this was still a long way from today's automated auctions, and early optimization models did not swiftly brush aside entrenched organizational routines and power relations by their sheer force and elegance. But the auctions and platforms that circulate advertising money and consumer data today would be impossible without the recurring efforts, started here, to configure media decisions such that the potential value of different advertising investments could be rapidly computed, compared, and discriminated using models and machines.

A critical implication of all this is that the OR/MS paradigm defined media as means for achieving advertising objectives and scored them with respect to their efficiency in organizing the attention of specific consumers. BBDO's LP model, for example, was not designed to cultivate an informed and compassionate citizenry or to help diverse and resilient communities govern themselves. Its explicit and only goal was to maximize rated exposure value. Many people consider it self-evident that advertising is necessary for democracy because it pays for journalism; yet these media selection techniques materialized the business's deeper commitments to treating media as differentiated investment opportunities and to optimizing those investments. BBDO's approach even implied that basing investment decisions on subjective perceptions about the qualities of a media outlet was irrational, and quantifiably so. As BBDO's media director explained in 1962, the LP method enabled the agency's planners "to arrive at a final media recommendation that concentrates on a marketing problem and not on the media themselves."[120]

There is irony and contradiction here. BBDO and clients like DuPont were known for "institutional advertising," making ads and sponsoring programs that showcased the morality of big business and American capitalism. Rather than producing sales, these sponsorships aimed to produce a certain form of citizenship and corporate governance, as well as an esteemed public image for companies or whole industries.[121] At the same time, these two firms stood at the forefront of advertising technoscience and its culture of optimization. This aspect of their identities fixated on the efficiency of marketing processes, measuring return on investment with a "profit yardstick."[122] But with both faces of their split personalities they practiced politics by other means. The political economy of US media bestows on advertisers a responsibility to judge and fund productions of public culture; taking advantage of computers and OR/MS, a part of the industry that executes those judgments learned to perform its responsibility as an objective, rational calculus centered on marketers' private values, while the industry overall remained passionately committed to reproducing what Zygmunt Bauman calls "a society of consumers."[123]

In another interesting irony, agencies sometimes stressed that optimization actually shifted media buyers' focus *away* from ratings "numbers" and toward audience qualities and "editorial values."[124] In other words, these tools directed media buyers to consider each audience segment's propensity to consume and each media outlet's effectiveness in accessing and influencing the most valuable people. *Don't look for the cheapest or largest audience; look for the combination of audience profile and media vehicle to satisfy the marketing objective*—that was the pitch. It was a handy device for agencies because it discouraged clients from slashing their spending and instead urged them to trust these technologies to squeeze more efficiency from their budgets. It is ironic, however, because modeling techniques quantified those editorial values and consumer profiles so that they could be calculated to optimize an objective function. These procedures flattened media into variable inputs for maximizing rated exposure or cost per prospect obtained.

While diverging from a simple "buy by the numbers" routine, the model-based rule of calculation aspired to enumerate *everything* that analysts deemed important to marketing objectives or decision situations. BBDO's research director admitted as much when responding to a question about whether LP might reduce the "ratings battles" among broadcasters. "I question very seriously whether we will ever be able to take out the numbers," he said. "However, we will probably break down audiences from sheer numbers into *kinds of people*: the times of day they watch, their demographic characteristics, the products they buy, etc. If anything, you will probably see a greater demand for numbers but of a more specific kind, more pertinent to the problem's solution."[125] OR/MS formatted consumers and advertising processes in such a way that, as measurement and computation expanded, transactions could be executed by way of numbers that signaled, for example, the probability of a certain behavior or the expected profitability of a potential customer. The initial maneuvers detailed here remained within a demographic mode of knowing and transacting, but they began to set up material and cultural conditions for imagining progress toward new ways of packaging commodified attention and behavior.

Overall, media models represented a "rationalization of audience understanding," to use Philip Napoli's phrase.[126] But it was a contradictory rationalization. Math men remade media and consumer audiences into properties that could be acted on with management technique and impersonal decision rules. But OR models also admitted more complex variables into the

calculations—variables that were difficult or sometimes impossible to measure reliably. By trying to solve one problem—accounting more precisely for
the value of advertising opportunities—the business created new logistical
and epistemological problems for itself. Optimization complicated media
selection, in comparison to simply buying exposures by ratings numbers,
and it created openings where subjective assumptions and judgments could
be reintroduced under the cover of a rational edifice. This contradiction
remains active in adtech. Programmatic advertising enacts and commodifies
rationalized units of audience attention or behavior via machinelike decision
systems, but those decisions accommodate many hard-to-know variables
and thus permit unaccountability to hide among seemingly objective facts
and figures.

CONCLUSION

Operations research and management science helped invent digital adtech.
Experts and techniques that responded to corporate pressure for efficiency
and rationality began to command legitimacy and power within advertising's organizational cultures. OR/MS specialists at consulting firms, large
corporations, major advertising agencies, computer and information service
providers, and prestigious universities set out to formalize marketing processes and relationships. Model building implied and manifested particular
orderings of the world—what was important, how things fit together, who
was worth what, which outcomes to expect, how to intervene and adapt.
And it required analysts and decision makers to assign explicit values to
whatever they wanted to predict and control. Market segments and media
vehicles were scored in terms of their value to advertisers, and these scores
affected the flow of money through ad-supported media industries. This culture of optimization set a lasting tone for adtech's discrimination engines
and created conditions for the eventual decoupling of media content and
advertising placements, a hallmark of programmatic advertising. Optimization models did not sever the link between media content and advertising
on their own, but they made concrete in new ways the desire to configure
advertising investments around marketing objectives.

Here, as in the previous chapter, we find technical experts trying to organize reality to make it calculable, predictable, and manageable via the know-
how they claim to command. The emergence of optimization models, like

other forms of rationalization, was about not only technical improvements but also a particular rhetoric of decision making and expert authority.[127] This mathematical elite acted like what Jenna Burrell and Marion Fourcade call today's "coding elite," vowing to eradicate irrationalities and inefficiencies from an industry by reforming its traditional habits through algorithmic designs.[128] Operations researchers framed media selection as a formalized science of decision making and translated media decisions into optimization problems with mathematical solutions. The management of advertising investments began to look like a sequence of information processing and analysis tasks that could be delegated to computer systems and the specialists who used them. Formalizing, automating, and optimizing decisions about advertising distribution are the essence of what adtech does today.

Though draped in dazzling ornaments of space-age technology, optimum seeking did not really upend the existing logic of media selection. Rather, it aimed to advance industrial practice closer to some of advertising's ideal purposes and priorities. Peter Drucker might have described this as a form of automation. The "basic automation question," Drucker explained in 1955, is this: "what is the logic of the process and how do we organize it according to its logic, rather than according to tradition."[129] Automation, here, is the recognition of an opportunity to refine or intensify the social relations contributing to some objective function. This is precisely what OR/MS claimed to be doing—aligning advertising investment decisions with their underlying logic. Optimization models would weed out the biases, inefficiencies, and irrationalities lurking in traditional (nonautomated) practices, revealing value that was ripe for the picking.

What Drucker calls automation is thus an affordance, the perception that a technology or technique can be used to organize a system or process more rationally. Caitlin Zaloom explains that this work of rationalization "is never quite finished: there is no end to the process of producing the conditions of formal rationality."[130] So it is with optimization. Optimization is a promise of improvement, and even as it identifies the *one best choice*, the promise is never finally satisfied. Every "disruption" implies the potential for a new optimum. Dreams of optimizing advertising are recurring dreams. They surface over and over, whenever changes, or even perceptions of change, invite well-positioned actors to claim that *now* it is possible to progress toward perfection.

II An Archaeology of Affordances

INTERLUDE

Business leaders, like leaders in other areas, are influenced by an "image of the future." Their ideas about where they are going may be clear or vague, but in any event they have a subtle and far-reaching impact on administrative thinking and decisions.

—Jay W. Forrester

Affordances are the possibilities certain actors perceive as being offered by an environment or resource.[1] With respect to technology, affordances exist in a relation between the material properties of an artifact or system and actors' recognition of what that technology allows them to do.[2] Peter Nagy and Gina Neff usefully emphasize that users and designers of technology often imagine possibilities that do not exist yet.[3] For our purposes, we can define affordances as the *situated recognition of potential*. Affordances therefore reflect and mediate the priorities, ambitions, expectations, and wherewithal of users trying to discern and activate a technology's capabilities—including "power users" and intermediaries with enough leverage or visibility to influence authoritative definitions of a technology. Institutional logics, power roles, and modes of accumulation organize these perceptions and the strategies for realizing them in action.[4] Affordances are materially grounded but socially shaped resources that actors use to structure conditions of possibility. They are a view of the future—of what *could* be—filtered through relations and dynamics at a certain place and time.[5]

This is a cultural and political-economic way of thinking about affordances. Rather than focusing on the psychology of how individuals experience objects, I emphasize how affordances are intertwined with strategic imaginaries and assertions of power related to what a technology means, what it could and should do, and who controls and benefits from it. Storytelling and politics are part and parcel of how industries and professional groups perceive a technology's potential to make or remake the world. Some of the most important things new technologies offer are narrative resources and opportunities to legitimize certain possibilities. Their existence affords an "image of the future," to use Forrester's terms. In this case, computers and other information technologies provided the advertising industry with magnetic and versatile characters for particular stories of progress. Advertising exercised a vivid imagination whenever even the prospect of technoscientific change made it look like the future was up for grabs.

The chapters in part II examine four interrelated affordances that advertisers, attention merchants, and marketing service providers have attached to information technologies and techniques: *programmability* (automation), *addressability* (discrimination and personalization), *shoppability* (interactive commerce), and *accountability* (measurement and analytics). These affordances are the building blocks of adtech. Their histories show that designs to optimize advertising were not side effects of the internet—they were actively cultivated across the second half of the twentieth century. Surveillance capitalism thrives online in part because companies have repeatedly hammered new technologies into the cracks in their marketing and management systems. Not only have they failed to plug the holes; because of all the hammering, the cracks have multiplied and spread.

5 HOW ADTECH GOT ITS SPOTS: COMPUTERS, AUTOMATION, AND THE ROOTS OF PROGRAMMATIC ADVERTISING

When *Billboard* magazine announced that "an era of 'Automation TV Buying'" had dawned on Madison Avenue, it reported that people in the industry expected "much of the guesswork and crystal gazing in TV" to be eliminated thanks to the "tremendous investment by the agencies and media firms" in digital technology. These companies enlisted computers to expand their capacities for managing information, coordinating transactions, and integrating complex calculations into media planning and buying. Automation promised to do more than just streamline clerical functions. A vice president at the Ted Bates agency welcomed profound advancements in technique, moving capabilities closer to underlying ambitions: "Not only will we be able to buy TV faster and more accurately than ever, but we may soon be able to relate television buys to the sales of individual products of clients and come up with rapid data on the sales effectiveness of TV on every station or network in the country." The article concluded with a comforting assurance that the "mechanization process" would only augment "the human aspect" of media buying. "No machine, all parties stressed, will ever replace sound judgement."[1]

This article was published in September 1957. If we overlook the fact that the "quick-thinking, multi-memoried business machines" described here were UNIVAC mainframes and that advertising executives marveled at the now-modest prospect of incorporating "as many as 30 or 40 factors" into media buying decisions, the report hardly looks out of place more than sixty years later. Programmatic advertising—using computers and other information technologies to automate and optimize media buying and selling—emerged more gradually than conventional wisdom admits. The arc of adtech's development looks different when we put programmatic advertising within a broader history of computerization. A leading historian of digital computers in American business says that advertising agencies were "some of the earliest users of the new technology."[2] Surprisingly, though, the history of computers and automation in advertising remains mostly unknown.

This chapter examines advertising automation from the 1950s into the 1970s. It focuses on the spot TV advertising market and showcases attempts to use computers to optimize media spending, to reform and speed up administrative processes, and to establish "online" interconnections among buyers, sellers, and service bureaus, enabling information to be circulated and transactions to be executed electronically and instantaneously. The industry discourses surrounding these efforts helped construct the meanings and uses of digital computers when they were "new media." These discourses illustrate that managing and exploiting data, and taming logistical and epistemological complexities, were already pressing concerns for the advertising industry in the broadcast era. In contrast to claims of a recent technological revolution,[3] this history reveals a more gradual *remediation* of advertising and media industries as important actors enlisted computers, databases, and communication networks to manage transactions and the information representing marketplace realities.

Inspired by David Beer's suggestion to examine how today's data analytics industry "cultivates a particular type of vision of data and its possibilities,"[4] this chapter details an early episode in which data processing and algorithmic decision making were socially shaped and positioned within imagined futures in advertising. Programmatic advertising represents an adaptation of digital technologies to the uneven but long-standing dream of accelerating and rationalizing the production of audiences and consumers. It embodies deep-seated desires to optimize allocative decisions and to account more precisely for return on investment. Amidst pressure to make marketing services more scientific, researchers and media directors at large advertising agencies linked their operations to computers' capacities to discern profit opportunities that manual analyses could not detect quickly enough. This construction of the computer's affordances helped advance the status of research and media departments, and it framed advertising problems in ways that called for mathematical and engineering solutions. The result was a sociotechnical reformatting of the advertising industry's spaces of calculation.

What seem like unique hallmarks of digital advertising figured conspicuously into advertisers' midcentury designs to take advantage of computing
and information resources. This insight adds new detail to Joseph Turow's
claim that the internet presented a favorable environment for media buyers and market researchers to exercise a set of strategies and skills developed
across several decades.[5] These strategies and skills, I argue, surfaced meaningfully in the remediation of spot TV advertising.[6]

A SPOT OF DEPARTURE

The term "spot" is often used casually to describe a discrete ad placement
on radio or television and to differentiate participating sponsorship (when
multiple advertisers pay to insert promotional messages, or spots, during
commercial breaks) from single sponsorship (when one advertiser finances
a whole program). More specifically, for our purposes, the spot market for
broadcast advertising refers to transactions in which advertisers or their agencies buy audience inventory from individual stations. Spot ads stands in contrast to the network ads that companies such as NBC distribute across a roster
of affiliates. In spot TV, units of available commercial time ("avails") are marketed either by local sales staffs or by sales representatives ("reps") acting
as stations' agents and dealing with media buyers for national and regional
advertisers. This chapter focuses on the national spot market, where advertising agencies buy audience commodities in specific market areas by engaging with independent rep firms as well as with reps selling spots for stations
owned by networks or chains. "National" here refers not to the distribution
of an advertisement but to the market territory covered by the advertiser.[7]

During the period under analysis, spot transactions typically involved a
sequence of relationships among stations, sales reps, advertising agencies,
and advertisers. After planning how to use a client's advertising budget to
achieve marketing objectives, an agency's media department submitted a
requisition for avails to one or more reps. Staff at the rep firm compiled a list
from pools of information about stations' inventory and transmitted it to
the agency. Buyers analyzed the list and assembled a purchase proposal,
which estimators assessed for expected costs. Once approved, an order was
placed and negotiated, followed by a series of communications to confirm
delivery of and payment for the bought audience. Transactions often concluded with the purchase of particular time slots, although negotiations

typically revolved around the acquisition of audience inventory at a stipulated cost per ratings point. The whole procedure activated numerous data retrieval, data input, and data processing operations; frictions arose from limitations in information storage and access, the speed and duplication of accounting and clerical work, and changes in avails, ratings, or other market variables. In sum, the management and analysis of information were crucial for buyers trying to make optimal decisions and for sellers trying to maximize revenue from their inventory.

The spot business can teach us a lot about adtech. Sensitive to time and geography, and involving many more buyers and sellers than network advertising, spot advertising was a testing ground for analytical and administrative tactics. Its signature challenges—such as distributing information about available inventory and coordinating commercial exchanges across dispersed actors—were logistical problems much like those addressed by innovations in online ad buying. Programmatic markets for digital impressions are spot markets, and some of the first efforts to establish the digital ad networks that scaled up internet advertising were launched by firms that specialized in selling broadcast spots.[8] Proponents called spot TV precise, flexible, and efficient. It permitted forms of geodemographic segmentation that roughly mapped onto spatial segregations of class, race, and lifestyle. A 1960 ad for one sales rep firm used vocabulary that is still familiar today, calling spot advertising "the only medium that allows you to reach—with maximum impact—the prospective customers you must reach, pinpointing only the markets you're in. Top advertisers get high return with low investment and there's no waste."[9]

Advertisers took these claims seriously. "What interests me most in a media department are the spot buyers," said the director of advertising at Pepsi-Cola. "The spot buyer in essence has a pile of 20 dollar bills sitting on his desk to dish out in exchange for fair return in the value of good spot availabilities. He is so important to us that we entrust millions of dollars to his buying discretion." Understandably, clients wanted spot buyers to execute their duties with care, exploiting whatever efficiencies their analytical tools could discern: "We expect them to rack their brains and spin their calculators to get a half thousand extra households through shrewd buying and constant improvement of schedules."[10] With growing demands to use demographic data for "pinpointing TV audiences," *Sponsor* declared in 1964 that "'computerization' of research" was "a trend now widely in effect in spot television buying."[11]

Participants in the spot market were not wholly successful at seizing the opportunities they perceived in computerization. But spot TV provides a useful case study because it illustrates the strain of holding together a complex and information-intensive market; thus, it was a fertile discursive space for asserting the progressive force of new technology. Without overstating its similarities to modern programmatic advertising, this early period of computer automation in spot TV lets us observe a moment when advertisers and their agencies seriously wondered how to take advantage of apparently revolutionary information technology. Their definitions of the situation manifested an acute desire to improve and accelerate calculative action and market administration. Even if the procedures seem prehistoric compared to today's real-time buying, they teach us something about how programmability was constructed as an affordance. With remarkable parallels to the automation evangelists in internet advertising, participants in spot TV envisioned computers as a means to refine existing procedures, exact new efficiencies, and steer advertising toward a more rational and machinelike system—to modernize flows of information and commerce.

COMPUTING, CALCULATING, OPTIMIZING

In 1962 an executive at J. Walter Thompson (JWT) described the computer as "a new tool for the human mind, enlarging its possibilities just as the telescope and microscope did."[12] JWT was among the first agencies to acquire a UNIVAC computer from Remington Rand and to announce plans to install a more advanced transistor-based data processing system. Preparations for the latter equipment—leased in 1963 from RCA, an agency client—were overseen by JWT's assistant treasurer, whose "primary job at Thompson [was] to study the flow of information and documents within departments, between departments, and to clients."[13] Beyond its immediate uses in accounting, computerization sustained hopes of better research, planning and forecasting, and decision making. Perhaps no application of data processing excited the business more than the potential to place ads in front of consumers more efficiently. Some of the earliest efforts to automate and optimize media planning and buying showcased what became durable cognitive and cultural frameworks for interpreting the potential of new information technologies.

"ELECTRONIC BRAIN FOR TIMEBUYING"

The American Research Bureau (ARB), an audience measurement firm that later became a subsidiary of a leading computer vendor, trumpeted an early contribution to "automated timebuying" in 1959.[14] ARB planned to install the newest UNIVAC model (figure 5.1), institute a more comprehensive data collection protocol (nationwide, county-by-county TV ratings), and establish rapid communications channels with advertising agencies. *Broadcasting* reported that this new electronic system "will automatically lay out a complete TV campaign." ARB's president, James W. Seiler, claimed that combining this computing power with the company's ambitious audience measurement system "will fulfill the timebuyer's dream of a complete information service. It will give timebuyers precise sets of facts to use in placing television advertising."[15] With the "data delights" provided by ARB's "electronic brain" (to use *Broadcasting*'s terms), audience manufacture could approach the market ideal of rational, calculative action supported by full information. "Agencies," Seiler promised, "will be able to buy spot TV on a completely logical

FIGURE 5.1
The American Research Bureau installed this UNIVAC magnetic amplifier solid-state computer in the late 1950s to expand its capacity to collect and manage audience information and to develop a system that could automatically plan a complete television advertising campaign. (*Source:* "Electronic Brain for Timebuying," *Broadcasting*, May 25, 1959, 32)

basis." Anticipation of UNIVAC's potential left the new personnel recruited to work with these resources "indulging in statistical fantasies."[16]

Over the next two years, ARB and its parent company, CEIR, adapted this system to suit BBDO's optimization model. As noted earlier, by 1961, BBDO had "developed, tested and placed into pilot use [a] computer process for selecting advertising media."[17] The agency used linear programming—an algorithmic approach—to sort through a set of important variables for media decisions. These variables included a "detailed 'profile' of the advertiser's customers," information about likely audiences for various media options, the availability and cost of commercial inventory, and the advertiser's budget. "With this information," one report explained, "the computer in minutes examines all possible combinations—which can run into the millions—[and] comes up with the one which, mathematically, best meets the advertiser's requirements."[18] Without computers, solving a linear program using the simplex algorithm involved a straightforward but time-intensive set of calculations. With digital computers, the optimum solution could be found in minutes. By the 1960s, IBM was offering access to programs running versions of the simplex algorithm through the company's computer time-sharing services.[19]

BBDO was a leading combatant in what *Broadcasting* called "a continuing battle on Madison Avenue to simplify by electronic means the highly complex and often highly subjective act of media selection."[20] BBDO initially hired CEIR as an outside computer service, but it soon leased its own machine from Honeywell, a client of the agency, at a cost of $100,000. BBDO's general manager boasted that the Honeywell installation "will mark the first case in advertising history where an advertising agency has totally integrated its marketing service operations with modern computer technology."[21] To automate media selection, all the relevant decision processes had to be carefully formalized. The goal was to eradicate subjective sources of inefficiency and thereby design a more rational calculative space. A manager from BBDO explained that the agency's computer "acts very much as a disciplinarian to force media planners to think in structured terms. The computer has no romance in its soul. It has no subjective opinions about things. Nobody can take the computer to Palm Springs for the weekend. Essentially the computer then forces us to put all of the ideas that go into it in terms of figures and facts."[22] According to the director of operations research at CEIR, the potential for using computers to optimize advertising plans and choices was limited only by the supply of suitable data. "This is a small glimpse at a future which can be as close as available information allows it to be."[23]

The early 1960s saw more and more agencies investing in "high-speed methods to improve the mathematical bases upon which media selection can be made."[24] By 1963, a quarter of the national television business was estimated to be "handled by so-called computer agencies."[25] A study released that year found that sixteen advertising agencies were "equipped for automatic data processing."[26] By another estimate, more than half "the major agencies" were "using or experimenting with computers in media evaluation and selection," as well as for routine administration.[27] Trade journals reported in the spring of 1964 that agency computer installations and usage had tripled over the previous ten months.[28] Two years later, an IBM salesman observed, "There's hardly an agency billing more than $20 million annually that doesn't have its own computer or use a computer service bureau."[29] (Thirty agencies spent at least $20 million on broadcasting alone that year.[30]) Surveys of the top fifty US agencies found that just over half the respondents in 1968, and more than 75 percent in 1973, had a computer or time-sharing facility at their offices.[31]

Along with BBDO, other major agencies setting the pace in digital computerization included Young & Rubicam (Y&R), Leo Burnett, and JWT. In 1960 these firms ranked first (JWT), second (Y&R), fifth (BBDO), and seventh (Leo Burnett) in broadcast billings, accounting for $389 million in combined spending.[32] By 1966, they were the top four spenders, doling out more than $720 million of clients' money.[33] These firms recognized the computer's ability to accelerate informational and commercial flows. A vice president at Y&R told an audience of broadcasters, "To us it has the positive advantage of doing tremendous quantities of analytical arithmetic with great speed and complete accuracy. It enables us to make better buys faster with fresher availabilities."[34]

Y&R's commitment to automation was perhaps the greatest among its early rivals. In 1962 the agency declared itself "in a race to the moon with our competition."[35] This sort of strategic escalation gripped the industry. By the mid-1960s, the consensus was that agencies required computers, yet they still operated the technology at a financial loss. The president of one rep firm characterized the agency situation uncharitably: "At this point they can't afford not to have them, they're important in attracting new business, but many agencies have expensive hardware that's being used for kindergarten purposes."[36]

Y&R's equipment was undeniably expensive. The agency spent $1 million to purchase a UNIVAC in 1960, and two years later it began leasing an IBM 1620 for $25,000 per year. By 1966, it was renting a Burroughs B5500 for $28,000 a month, under the supervision of a "data and systems" division created two years earlier.[37] Agencies in general reported spending around $50,000 to assemble data for computers to analyze, and according to media directors, the average cost of having a machine answer one question, or the "cost per computer run," was roughly $1,000.[38]

Y&R was also committed to graduating beyond kindergarten. Designers at the agency developed a computerized selection system around an iterative algorithm called the high-assay model (figure 5.2). One observer called it "the most ambitious of all media models." It could prescribe not only where

FIGURE 5.2

Designers of Y&R's high-assay model. Personnel fed data into the computer about the product's "behavior," the media options, and the advertising, along with a mathematical decision system. The machine then returned a schedule optimizing reach, frequency, and periodicity (i.e., when to advertise). Martin Starr of Columbia University is seated at the controls. (*Source:* "Y&R, BBDO Unleash Media Computerization," *Advertising Age,* October 1, 1962, 118)

and how much to advertise but also when. Furthermore, "It can take as its objective the result of the schedule in terms of brand share, i.e. it uses figures for advertising effectiveness which few practitioners claim to know."[39] The agency was also using its computer to fabricate synthetic information with its Data Breeder system, imitating Simulmatics. With this system, programmed for Y&R's UNIVAC by Remington Rand, the agency could "fill in the millions of empty spaces [in a database] if we know only 10 or 12 things about our audience."[40]

By the mid-1960s, more and more agencies were imitating or extending these computerization efforts. McCann-Erickson developed "a highly sophisticated and completely computerized system for comparative media analysis and allocation of advertising investment."[41] In 1964 seven major agencies formed a consortium and pooled their resources to help them keep up in this expensive race. They hired the Diebold Group, a consultancy that specialized in automation and information technology, to act as "management engineers," "mathematical consultants," and computer programmers for an automated media planning system. As *Sponsor* reported, the "basic approach" resembled what the top agencies were doing: "[it] makes use of simulation and incorporates an 'optimizing' procedure designed to search out the best allocation of advertising budgets."[42] Three more agencies had joined the initiative by 1966, working together on what they called the "computer optimal media planning and selection system."[43]

The steep costs of equipment and information gathering created both pressures and opportunities for the professionals claiming authority over and through computers. According to JWT's media director, "The stake the media department has in this venture has given it a new identity in the eyes of agency management." Media professionals faced the possibility of either advancing their status or having their functions "absorbed into some new super-department presided over by mathematicians and researchers."[44] Computerization's promise and its price tag thus had performative effects. "The heat is on to produce evidence for agency management that their huge investments in computers and research are paying off in terms of new and productive planning techniques."[45]

Much like the general corporate rush to develop management information systems, it was not clear that agencies' investments were paying off. Technoscience served, at least in part, to attract or impress clients—including the computer manufacturers themselves, which were becoming major advertisers in their own right, trying to outmaneuver competitors and define a new technology.[46] With many large advertisers already well invested in computerization and operations research, agencies worried that if they didn't promise capabilities exceeding what clients could do themselves, those clients might "succumb to the lures of setting up a house agency."[47] Advertising agencies publicized their computer installations to brand themselves as sophisticated and futuristic, and computer companies relished the flattering hype.[48] Yet, for all the apparent majesty of optimization models, one had to squint hard to see them delivering on their promises. An operations researcher diagnosed the problem as he saw it in 1962: "What we are all lacking is a coherent, significant and indisputable mathematical theory of advertising response. Advertising is where chemistry was in the 14th century—in the alchemist's cell, with the dried frogs-legs and the magical incantations. . . . Without a quantitative mathematical theory to guide its application, a computer—except for trivial data-sorting and arithmetical operations—is useless. Frogs-legs and incantations are cheaper, and possibly just as effective."[49] In other words, advertising needed an understanding of behavioral influence that was as muscular as its computing tools. Those things were entangled, of course; as computation spread throughout the organizational and intellectual cultures of advertising, ways of understanding consumers were materially and strategically linked to the techniques for assembling and ordering market information. Still, many saw untapped potential. As the president of the Marsteller agency put it, "We are just beginning to 'scratch the surface' of electronic data processing applications in marketing. . . . Innovations and potential possibilities are developed practically daily."[50]

Space-age optimization models may have made data sorting and arithmetical operations seem humdrum, but the latter were hardly trivial to media buying and selling. While futuristic visions helped set a path for development, the buyers and sellers of TV audiences were actively using computers to confront more immediate priorities. Throughout the 1950s and 1960, agencies and sales reps worked at automating administrative and transactional routines, incorporating new information technologies into logistical utilities for managing spot TV.

COORDINATING THE SPOT MARKET

The commercial media environment was growing in scale and complexity in the middle of the century. In 1950 there were 107 commercial TV stations reporting revenue that totaled just over $1 million. By 1965, the 588 stations broadcasting throughout the United States accounted for almost $2 billion in revenue.[51] Particularly with the increase in UHF stations, observers noted that "audience fragmentation," as one agency executive called it in 1964, would complicate the job of spot buying.[52] To reach the people spread across all these stations, buyers had to maintain relationships with more trading partners. The number of sales organizations representing stations to national and regional advertisers nearly doubled from 66 in 1950 to 130 in 1960.[53] Added to this extensive development of the market were intensive challenges. Beginning in the 1950s, the transition toward participating sponsorship—whereby advertisers bought interstitial commercial slots rather than financing an entire program—meant that advertising inventory was divided into smaller units and sold to a wider range of buyers.[54] One study found that the number of prime-time programs on the three TV networks sponsored by multiple advertisers rose from just ten in the 1955–1956 season to fifty-seven in 1964–1965.[55] While this meant that an advertiser could distribute financial risk across more commercial placements, it also meant that buyers had to sort through many more placement options. Buying those slots was dramatically cheaper than sponsoring a whole program, but prices were still significant and rising rapidly.

Already by 1957, television was considered "the most complex and paperwork ridden advertising media."[56] Television networks and station owners had invested in data processing in the 1950s, mostly for accounting and billing functions.[57] But as networks looked forward to "harnessing [the] full capacity" of these machines, managing information and coordinating transactions with advertisers registered as pressing issues. "The computer's advantages as a sales tool to pick and chose [sic] between increasingly fragmented network inventories for optimal sponsor benefit are clear," *Broadcasting* reported in 1966.[58] As a sales director at NBC admitted, "Data Processing is a competitive way of life. Today's executive must accept and understand it."[59]

Observers saw special potential for automation in the planning, buying, and selling of "more flexible media" such as spot TV.[60] Purchasing spot TV presented the "most difficult media buying job," in the opinion of one agency executive.[61] The associate media director at Grey Advertising reported in 1959 that "five or six times as many man-hours may go into investing a spot dollar as a network dollar."[62] Paperwork and transaction costs were formidable, and the whole business hinged on rapid communications. The spot trade activated what one sales executive called a "constant churning of information."[63] Leo Burnett's vice president of media reported that his agency bought 350,000 broadcast avails in 1962, of which 160,000 were TV spots. He estimated that on a monthly basis this activity generated 1,500 TV contracts and at least 6,000 pages of buying estimates.[64] The president of Data Communications Corporation, a technology supplier to the industry, noticed that "the business automation systems that most resemble the requirements of broadcast are those developed by the airline companies for reservations." In fact, according to *Broadcasting*, spot advertising was even more complex.[65]

The administrative burden of processing transactions in this market escalated throughout the 1960s, taxing the capacities of existing systems and protocols. National nonnetwork TV time sales ballooned from $345 million in 1958 to nearly $1 billion in 1968.[66] By 1970, and not for the first time, spot sales to national advertisers ($1.10 billion) and local advertisers ($589 million) totaled more than half of television ad revenue ($3.24 billion).[67] The aggregate volume of inventory managed by US TV stations was becoming overwhelming, estimated at roughly twenty-three million advertising spots in 1967 and as many as thirty million in 1968.[68] As one source acknowledged, "The problem of accurate accounting between agencies, advertisers, stations and performers has increased in direct proportion to commercial volume, to the point where it is inadequately handled by procedures largely carried over from old radio days."[69] The sales side often had to wait 90 to 120 days to close accounts receivable, and agencies struggled to document errors and negotiate make-goods (i.e., compensation for a station's failure to run an ad as scheduled or to deliver the promised audience). The media manager at Benton & Bowles admitted that for a routine campaign, which involved buying spots across 132 stations, the agency experienced discrepancies between its purchase orders and the affidavits provided by stations on 32 percent of billings.[70] Broadcast Advertising Reports, an auditing service that monitored whether TV stations inserted advertisements according to agreed-on specifications, estimated that one large advertiser it worked for could expect as many as twenty thousand "irregularities" in 1963.[71] A

decade later, an advertisement for a firm marketing transaction processing services alleged that "the broadcast advertising industry is paying
$50,000,000—mostly in unaccounted for expense—to reconcile discrepant
paperwork and payments."[72]

Advertisers complained about being tangled in a "jungle" of paper. The
manager for media buying at Gillette said one the "major goals" of his organization "is to deliver the right message to the right audience at the right
time," but "achieving that goal can involve unlimited hours of paperwork."[73]
These difficulties left *Broadcasting* wondering: "Can automation organize the
buying and selling of broadcast advertising? Nobody knows yet, but lots of
people and hardware are at work on it."[74]

CIRCULATING SPOT TV'S LIFEBLOOD

The needs of clients drove agencies to invest in data processing systems.
For example, Y&R was one of fourteen agencies servicing Bristol-Myers, a
leading marketer of pharmaceuticals and consumer packaged goods with
a $130 million advertising budget that included roughly 300,000 TV spots in
1966. "Just to keep track of all the television time it buys, on all the stations it
uses," *Fortune* reported, "the company needs several computers. One, owned
by Young & Rubicam, is the heart of a tele-type-linked network of agencies
reporting on Bristol-Myers' daily purchases of TV station time around the
country."[75] Bristol-Myers was not in a league of its own. The $25.5 million
it spent on spot TV was only good enough for fifth on the list of top spot
buyers in 1966—and just 1 percent ahead of sixth-place Lever Brothers.[76]

Even before the dramatic growth of TV, agencies and reps sought ways to
contend with spot advertising's administrative demands. By 1955, Leo Burnett had "entered automation . . . to attack the paper problem."[77] The company installed an IBM 305 in 1960,[78] and less than two years later it began
leasing an IBM 1401 system that "automate[d] much of the clerical functions of television and radio spot buying." The computer system consolidated the generation and printing of station contracts, ad schedules, and
invoices for clients. Computers were especially useful for calculating prices
and factoring in discounts that applied at certain volume or frequency
thresholds. Without a machine storing this information and automatically
adjusting rates, estimating the relative costs of different schedules was a
tedious process. Leo Burnett's media director boasted, "in the old days we
needed several weeks to prepare a detailed broadcast estimate for a client.

Today it's a matter of minutes or seconds for computation and printing by our programmed computer method."[79]

Computers reportedly halved the amount of "handwritten data" required of an agency's buyers and estimators.[80] The vice president of an agency that managed $10 million in billings, mostly in spot broadcasting, claimed that the use of computers allowed his firm "to compress the entire buying procedure" into eleven working days, down from the twenty-nine days typically required to process a transaction manually.[81] Another agency adapted a computerized system to better organize the market information presented to its timebuyers, making the relative value of each avail easier to discern. "The program we finally put together," the agency's general manager explained, "is designed to bypass the endless hours of manual computations and put in our buyers' hands a simple ranking of television programs and radio time periods that fall in descending order from the most desirable to the least desirable for any given product."[82] The spirit of these designs is similar to modern campaign-management dashboards that digest a glut of information into easily-interpreted metrics and graphical guides. With market realities represented as rationalized ordinal rankings, decision making becomes programmatic.

Spot TV sellers were also preoccupied with processing information. The rep firm Peters, Griffin, Woodward (PGW) developed a system around a UNIVAC machine in 1957 (figure 5.3). Information about availabilities, audience ratings, order confirmation, schedule changes, and a range of statistics about media markets constituted "life-blood matters" in the spot sales business.[83] Echoing theories of an ideal market, PGW installed its "robot genius" to "meet the demands for complete and accurate information."[84] The system was designed to provide this information to prospective buyers at a fraction of the time and effort previously required. PGW planned to "deliver availability reports in about one-sixth the time that it takes a trained secretary and salesman to do the same job."[85] For example, the company claimed its machine could print one hundred lines of availabilities per minute, compared with the roughly thirty minutes needed to complete an average fifteen-line availability sheet.[86] Later, however, PGW complained that the benefits of using a computer to report avails to buyers was diminished by the tedium and cost of reprogramming it to keep up with fast-changing information.[87]

Other large rep firms followed PGW's lead. By 1966, H-R was spending $300,000 on yearly computer-related expenses, and it was poised to install a

FIGURE 5.3

Women working as "coordinators" in PGW's system. In response to avails requests routed by sales staff to the data processing center, the coordinator retrieved the desired information from the master reference files stored on a Wheeldex machine (pictured on the left) and produced a punch card that went to the "expediter" for processing. (*Source:* "Automation Speeds Spot Sales," *Broadcasting-Telecasting,* October 7, 1957, 72)

third-generation (integrated circuit) system, the IBM 360. Its investment in data processing was designed to accelerate and increase control over paperwork processing and to centralize billing operations for the many stations it represented.[88] In comparison to manual data input and retrieval, the IBM 360's "random access" storage facilitated "a search for information at fantastic speeds."[89] Illustrating the centrality of information flows to the spot TV business, H-R reorganized its corporate space "so that offices radiate out from the computer." Not to be outdone, the rival Katz Agency planned to bolster its $100,000 annual outlay on computer program development with another $400,000 for data processing personnel.[90] A vice president at Katz "hailed the advent of the 'machine' and chided doubters as standing in the way of more efficient time buying and selling and more dollars for spot television and radio."[91] Katz had invested in an IBM machine and consulting services from John Diebold Associates "to equip [its] television sales staff to meet the most exacting demands of the future."[92] Before long, the Blair company joined the list of national rep firms that had "recently installed immensely sophisticated systems for both radio and television which supply their salesman [sic] with dozens of availability, sales and demographic reports that, it would seem, all but eliminate the need to sit behind a book filling out forms."[93]

Adjacent businesses found ways to insert themselves into spot advertising's circuits. A sector of intermediaries emerged to handle elements of transaction processing. Acting as "automated billing houses," Central Media Bureau, Broadcast Clearing House, and Broadcast Billing Company used electronic computers to "speed the placement and payment of spot broadcast advertising for agencies, stations and station representatives." Other firms marketed a variety of data and research services, emphasizing the competitive necessity of strategic intelligence. Subsidiaries of large technology and telecommunications companies also rented their computer facilities to agencies and advertisers seeking more power to analyze information related to media selection.[94]

Given the potential to accelerate the pace of spot markets, BBDO's media director expected that "in five years [from 1963], computer usage will be as common to good business as the typewriter."[95] Like earlier developments in bureaucratic technologies, the growing supply of data processing power in the ad industry did not satisfy demands for information and its management.[96] Instead, demand for and supply of data resources reinforced and reshaped each other. Domestic and international expansion at major advertising agencies deepened their reliance on communications facilities to execute standard operations at greater scale, and these resources also opened possibilities for firms to render new services. When JWT ordered an RCA 301 computer system in 1962, its vice president of finance commented in the company's newsletter on the significance of this investment: "Recent new developments in equipment and approach have made it apparent that electronic data processing within JWT is not only feasible but is also, in view of our increased growth, essential for providing a range of information and services that is already or will soon become necessary in our business."[97]

If today's advertising business is driven *by* data, in the 1960s it was driven *for* data. "Mindful of the computer's voracious appetite for data," *Advertising Age* reported in December 1962, "advertisers are frantically beating the bushes for better information on who their customers are, and where and when they buy. Even dusty product warranty cards are being hauled out of warehouses and examined for possible information."[98] BBDO complained that its computer was "starving for more data" about audiences and advertising effects, and the agency was reportedly paying as much as $50,000 to cross-tabulate figures from A. C. Nielsen, Daniel Starch, and other market and media researchers.[99] From roughly 1961 to 1966, BBDO spent more than $1.2 million buying information about media audiences.[100] The vice president and director of marketing services at Foote, Cone & Belding admitted that "the use of computers 'has sparked an intensified demand' for information in advertising and media."[101] In 1964 a representative from Leo Burnett claimed that the firm spent $250,000 annually to buy "raw data" for research purposes, plus another $130,000 for the "'machinery' to process it."[102] Two years earlier, he had told a gathering of the American Association of Advertising Agencies (4As) that to make computers truly useful in media selection, "we need far more and far more accurate media data than we are now getting."[103] Stations, rep firms, and media researchers all noted "the explosion in demographic data now demanded by ad agencies."[104] As an executive at a leading rep firm later recalled, "research that was being demanded by the advertiser and the agencies forced companies' research departments to become more sophisticated. You needed more research to analyze the data available and to create new ways of measuring data."[105] In a 1963 speech to the Chicago Broadcast Advertising Club, a manager from BBDO issued a litany of ways broadcasters needed to accommodate computerization in their business routines. As *Variety* summarized, the demands began with numerical considerations: "Be ready to provide consistent and accurate data on size of audience and its components for the machine to use, or be shut out of consideration."[106]

Not all parties welcomed the demand for data—namely, the publishers and stations from which agencies demanded it. "The data vacuum has hit some media right in their pocketbooks," one journalist remarked.[107] The president of a station ownership group complained at the 4As annual meeting, "No medium has provided its agencies and advertiser customers with as much 'buying' information as broadcasters. We have researched ourselves almost to death, and your new electronic pets, the computers, threaten to finish us off."[108] He added a prescient lampoon of the direction data-driven behavioral targeting would take: "Red-haired housewives between 18 and 39 with three children who do their ironing on Wednesday mornings may be the best sales targets for a particular product, yet the task of determining the media facts necessary to pinpoint the target may not be worth the cost."[109] The frenzy to assemble those facts only churned faster, however, despite frustrations and warnings. Today's online publishers face similar pressures to install tracking code on their sites so that adtech companies can measure and optimize advertising events.[110]

Information has always been the lifeblood of the spot TV market, and from the 1950s onward, computers were adapted to meet buyers' and sellers' demands for both timely information and the resources to deal with so much data. A media executive from Benton & Bowles testified to this point in discussing the "on-line computer system" his agency was instituting in 1970. "The system has been built," he admitted, "for the specific purpose of enabling us to handle more efficiently spot TV's complexities."[111] He was not alone in recognizing "the inherent complexity of the spot-TV business—particularly as it relates to the communication and paperwork processes among buyer, seller and station." And he identified these difficulties as critical to motivating larger technological and organizational changes: "that complexity has helped spawn the development of outside buying services, electronic monitoring systems and firms specializing in computerized post-buying analysis, as well as the development of standardized availability forms, contracts, invoices and billing cycles."[112] Computers were incorporated in these efforts to install a more machinelike system for managing transactions in spot advertising. And, in a process of mutual accommodation, the business had to be formatted to suit the machines.

FORMATTING A MARKET

Computerization affected not just the volume of data but also their format. "The computer age will 'implore' media to seek more data, and to make data uniform," *Advertising Age* suggested.[113] There were contradictions to manage here. Information technologies promised new possibilities for market segmentation, yet those new segments had to be operationalized as objects suitable for routine business. Publicizing a 1963 report by the 4As, *Broadcasting* noted, "There was no question that the need for an attempt to set up a guide or standards for data on audience types has been hastened by the computer age."[114]

One early response to these challenges illustrates the importance of building sociotechnical scaffolding for marketplaces to support calculative action and durable relationships. In 1966 the Television Bureau of Advertising

(TvB), a trade association representing commercial broadcasters, announced that it was budgeting $300,000 over the next five years for "liaison work" to help agencies, stations, and reps cohere around standardized information and communication protocols, which many considered a prerequisite for "industrywide integration of computerized buying and selling procedures."[115] The next year, TvB touted a "blueprint for a uniform, computer-based system of spot television buying that could revolutionize the business."[116] The system of spot (SOS) was designed to streamline administration and make spot TV easier to buy and sell. SOS introduced new paper forms to create uniformity, reduce duplication of clerical work, and assign consistent codes for classifying inventory units and types of transactions. Most important, the forms were designed to format information and commerce in ways that supported the "ultimate plan to computerize the buying and selling functions."[117] The manager of the project claimed that replacing the various forms and conventions in use across agencies, stations, and reps with common, codified procedures for requesting avails and confirming orders would rationalize the presentation of information—and, indeed, the representation of the market itself—to buyers and sellers. The information representing the market could then be processed and exchanged among machines. Even as this initiative progressed more slowly than hoped, TvB remained committed to improving connectivity and eliminating friction from commercial and informational flows. "The idea is to make it easier for everyone to talk to one another. The goal: to make spot advertising easier to buy."[118]

These materials and systems for recording and categorizing market activity reflected plans to make advertising transactions programmable. Programming implies an effort to codify procedures and prescribe outcomes, but more generally, programming can refer to a process of assembling and organizing.[119] Information and communication technologies, including mundane tools such as paper forms and alphanumeric codes for identifying elements in these transactions, were applied in the 1960s to lubricate commerce—to make it easier to buy and sell audiences. Like later developments in computer spreadsheets and software for managing inventory and expenditures, the SOS introduced new resources for rendering, analyzing, and ordering the elements that make up this market.

Logistical technologies of this sort configure actors and relationships in ways that produce specific possibilities. They have been essential for efforts to accelerate market exchange and systematize the retrieval, routing, and documentation of advertisements. Automation enthusiasts imagined that once the information coursing through the spot market was fed into frequently updated databases, it would be possible to interconnect those databases and orchestrate transactions electronically. "Eventually," one report confirmed, "the plan envisions a wired network linking agencies, reps and stations with a central computer into which orders would be fed and which would then make all the necessary calculations and automatically feed the information back to the agencies, reps and stations."[120] As the next section shows, TvB was not alone in imagining a future of interconnected machines and automated calculative action.

VIRTUALLY INSTANTANEOUS: NETWORKED EXCHANGE

As agencies and reps invested in computerization, many saw the potential for networked interconnections between buyers and sellers. In 1963 an executive from Leo Burnett mused, "I think it's very reasonable to assume that the rep and the agency will directly tie-in into one another's installations" to immediately access information about availabilities.[121] By "linking" computer systems, he argued, the industry could automate "as much as possible the entire TV spot ordering, buying, billing, paying, reporting operations."[122] Y&R's director of computer applications shared this enthusiasm: "The potential for combining agency use of computers in processing spot availabilities and purchases with computer programs operated by rep firms would appear to be almost limitless. The result should be greater efficiency, greater accuracy, and very importantly, greater speed in processing that highly perishable commodity—spot availabilities."[123]

Before the end of the 1960s, some of the largest agencies had established direct data communication with rep firms via dedicated telephone lines. The director of media and research at one agency explained how this affected daily routines: "Each evening the agency computer queries the reps' computers for the latest availabilities in a given set of markets. Then, by tapes of ARB or Nielsen data, the agency computer can complete the rating and cost-per-thousand information the buyer requires. Each night a printout is prepared and placed on the buyer's desk for immediate action the next morning." Compared with the chains of interactions previously required to obtain and organize this information, he claimed, "this method employing computers is virtually instantaneous."[124]

Some commentators were careful not to get swept up in futurism. Even proponents admitted that an "imaginary miracle machine" capable of automating the entire process of media buying was "not yet on the horizon" in the mid-1960s.[125] *Broadcasting* observed in 1966, "The histories of companies born to supply the station-rep-agency triangle with automated central billing are replete with disappointments or, more to the point, failures."[126] Still, in spite of skepticism and anxiety about human replacement, many observers looked forward to a future when more media planning and buying would be handled by programmed machines. Dan O'Neill, president of the Advertising Data Processing Association and a researcher at McCann-Erickson, was one of what *Broadcasting* called "a growing school of media men who are trying to turn mechanical media selection into a more reliable science."[127] O'Neill predicted that "the mechanical revolution" would result in "fewer and larger rep houses in the future, all eventually working on an integrated real-time computerized system." Others imagined that the "ultimate in computerized inventory control" would be "an on-line representation system [in which] all sales offices and stations would be connected by wire to a central computer, feeding sales information as quickly as it becomes known." Foreshadowing later developments in self-service dashboards and trading desk software for internet advertising, the president at H-R expected that, by the early 1970s, such a system would allow a salesman to "punch a keyboard at his desk with availability requirements and get an immediate visual response from the computer on a TV monitor."[128] Looking back on rep firms' computer plans, *Broadcasting* concluded, "in general they tended to envision the machines as providing virtually instantaneous links to stations and agencies and serving as storehouses for avails that the rep could sell faster, in more quantity than ever before."[129] Discussing a similar "console" spot buying system Benton & Bowles would implement over the next few years, one of its vice presidents looked forward from 1967: "The computer will give the agency media buyer instant access to spot availabilities in every market. With a simple desk console he will be able to check availabilities and buy spots—without ever using a pencil or picking up a phone." He expected that by 1972, "the entire spot-buying process—all the way from order, to conformation, to bill, to final payment—will be computerized."[130]

Spot buying and selling have long resembled financial trading and information systems. At a 1973 workshop convened by the Association of National Advertisers, participants hoped computers could be used to create "a kind of stock-market approach" to buying spot TV.[131] Within a few months, *Broadcasting* reported that the two leading providers of "on-line computerized information systems" to the commercial television industry had undertaken construction of a "three-way computer tie-in linking stations, ad agencies and rep firms." Data Communications Corporation operated an online system for sixty-seven radio and television stations, and Donovan Data Systems provided a comparable service to fourteen of the top twenty ad agencies in the United States. With the top three rep firms committed to participating, *Broadcasting* called it "the first step in an 'industry-wide common use of computers' to handle national- and local-spot accounting 'from the point of buy right through to the final payment of the invoice.'"[132] By January 1976, the Katz Agency, which was not included in that group, claimed to be the "first independent station rep to operate [an] on-line availability-retrieval system," which linked twelve sales offices to its New York computer facility.[133]

In many ways, these were primitive steps toward the programmatic supply- and demand-side platforms that interconnect digital advertising markets. Advertising networks like DoubleClick essentially absorbed computerized repping functions for online publishers, and supply-side platforms actualized these forecasts of consolidated, electronic avail depots, providing access to a constant stream of inventory. The envisioning of a programmatic future indicates that even if the technical ability was barely under construction, the desire and demand to manage more data at greater speeds existed at the dawn of the computer age. Even then, generating, processing, and coordinating flows of information and commerce were the bedrock of audience manufacture. In these early efforts to cope with mounting administrative pressures and to produce more rapid and rational decision making, we glimpse the underlying dynamics that continue to motivate developments toward advertising automation.

But as computing resources were appropriated to service existing demands, expectations changed. No doubt hoping to showcase his company's products, a representative from RCA urged advertisers, agencies, and reps to take "initiative" in testing the computer's capabilities. Suggesting that they dream up exotic questions to ask the new machines, he assured them, "computers can provide more answers than we have intelligent inquiries."[134] The implication, borne out as researchers attempted more sophisticated forms of statistical inference, was that new information technologies enabled new ways of seeing, knowing, and managing reality. Practical consequences were experienced at the organizational level, as agencies such as Campbell-Ewald reoriented office spaces, shuffled personnel, and adjusted routines and responsibilities within and across corporate departments so as to "tie in with the hot computer system."[135] Executives at Needham, Harper & Steers noted "the unlimited potentials of the computer for media research and planning." The agency's president declared that the value of the computer was constrained "only by the imagination of the people who plan its use."[136]

One nagging concern, though, was whether these plans included people. The legitimacy of automation technologies and algorithmic judgments is mostly taken for granted in advertising today. But trust in computers was not a natural endowment for everyone.

A FUTURE FREE OF FRICTION (AND PEOPLE?)

Despite considerable excitement surrounding computers, appreciation for these "high speed electronic wizards," as one media manager called them, was not universal.[137] Automation technologies were interpreted variously as levers of advantage or sources of anxiety, and almost always they signified a future up for grabs (figure 5.4). Ad agencies accommodated computers relatively quickly. The supply side of the audience marketplace was more resistant. Reps worried that their jobs might be deskilled or eliminated entirely. In 1958 a vice president at H-R cautioned against simply feeding ratings data into an electromechanical decision engine. "If we continue to play at the numbers game," he warned, "we may all be replaced by a single Univac machine."[138] Although most doubted this eventuality, expert opinion was not always comforting. "While there are still a lot of problems to be solved," a consultant to Leo Burnett speculated, "it is only a matter of time before any piece of information about any advertising medium will be obtained instantaneously without having to go through a salesman intermediary. There will no longer be a need to have the media salesman per se on the other side of the desk."[139] Other "prophets" predicted "the total eradication of repping in favor of rows of buttons and computer displays."[140]

Broadcasters were also wary. As one observer said about the prospect of a centralized online exchange, "Stations aren't going to put themselves at the mercy of the machine. They'd go bankrupt. And no good station manager lets all his inventory out of his control."[141] Throughout the history of automation, the supply side has feared a "nightmare of lowball rates submitted

FIGURE 5.4

Computers possessed a range of capabilities, empowering or threatening various professional groups. This cartoon depicts how different groups perceived computers' impact on their status and autonomy. Notably, "The Media Director" exercises control over dutiful servants, while "Media Salesmen" scramble to satisfy an intemperate baby. "Management" interpreted computers as fortune-tellers, while "The Marketing Boss" hoped to examine consumers through a magnifying lens. (*Source: Advertising Age*, December 17, 1962) to reps by buying scavengers."[142] Recent developments in programmatic advertising have clearly vindicated those fears.

Automation anxieties also went deeper than economics. Some station managers exhibited "psychological resistance" to computers. Fearing "they would lose their jobs or be reduced to mechanical handmaidens," they "weren't quite prepared to turn over the nuts and bolts of their livelihood to a piece of blinking hardware."[143] An ad for the Cosmos Broadcasting Corporation made a full-throated plea for humanity: "In these days of audience delivery based on computer-analyzed demographics, cloned programs, media buying untouched by human minds, computer-controlled this and automatic by-the-numbers that, we'd like to express a few thoughts about the human equation. . . . Computers are our tools, but until we find one that can say 'ouch' or 'wow' we'll keep striving to build warm human relationships with all those to whom we hold ourselves responsible."[144] The tone of this copy suggests that by the end of the 1970s, human relationships were perceived to be under assault, if not already subjugated.

These themes also turned up in discussions on the buying side, including doubts about computers' capacities and concerns that the "monster in the air conditioned room" would make personnel redundant.[145] One media director worried in 1955 that the job of media buying would be "reduced to statistical brainwashing." He criticized his counterparts at other agencies for committing to "the Univac timebuying approach," with its ruthless focus on cost efficiency. The "next move" for the data-driven media manager, he joked, "is to go to 57th St. and Madison Ave. and shop for his timebuyers at IBM World Headquarters. After all, IBM machines can add faster, compute faster, and get a lower cost-per-thousand faster than any human being."[146] Even an executive at the computerized Y&R agency reacted against this swelling tide of numeracy, complaining in 1961, "The moment we bet our all on one-tenth of a rating point or on a five-cent difference in cost-per-thousand, then we have departed from reality."[147] A year later, another skeptical media director pictured "all-out mathematical selection" as an existential threat. "No matter how exciting it may seem at present, it's not the present that's at stake. It's the future. If computers right now can do everything but judge, what happens when they learn to judge as well? I thing [sic] we've built a monster to destroy us."[148]

For the most part, though, agencies acted as a welcoming committee for the wonders of the information age. Of course, they rarely let computers take all the credit. Even conversations glorifying computer magic usually ended with tributes to the sanctity of human judgment. When asked whether computers would replace media buyers, a researcher at Grey Advertising replied, "Not until they can make one that thinks. It's still a machine. It will probably be a terrific step forward but it can't know all the intricate problems or the details of coverage or all the other things a timebuyer must know."[149] This sanguine attitude helped disarm fears of workforce redundancy, and it also implied that each agency's personnel possessed unique expertise that rivals could not replicate with a machine. This was part of a concerted effort to reinforce the legitimacy of automation. Y&R circulated an instructional pamphlet, *How the Elephant Bought His Spots*, to inform readers how its computer helped the agency make decisions. The machine "does not supersede a media buyer's judgement," the pamphlet promised, but supports buyers by doing "tremendous quantities of analytical arithmetic with unparalleled speed and chilling accuracy."[150] A BBDO memorandum defended against worker alienation by calling its linear programming system "an effective tool for the intelligent, creative media man, not his replacement."[151] William Baumol of Princeton University added academic authority to such claims, maintaining that the computer would be an "efficient assistant," but "ultimate decisions and judgements will always remain in human hands."[152]

Agencies issued these soothing statements to calm anxieties about technological change that they had stirred up by touting the revolutionary power of their computer systems. Not all agency managers made apologies about welcoming a more machinelike configuration. Some considered it inevitable that media buying would be highly automated in the future. The vice president of media and research at Papert, Koenig, Lois expected machines to be buying ads by the late 1960s. "Computers will buy spots, will select among network opportunities, and will do it better than the flesh-and-bloodniks," he wrote in 1964. Regarding the implications for existing workers, "one thing is certain," he claimed: "They can't compete with Mr. Machine. They've been technologically displaced." Although he predicted staffing reductions at media departments, he also thought buyers would absorb the responsibility to "harness and direct this large-scale analysis mechanism." He envisioned a symbiosis of worker and computer, with the buyer "constantly feeding and reading the machine." Unlike those who derided this as robotic button pushing, he expected the tight relationship with computers to elevate technically skilled media professionals into higher managerial roles in advertising organizations.[153] Others agreed. "The way we see it," wrote the media director at Fuller & Smith & Ross, "EDP [electronic data processing] is the final catalyst in the media man's evolution" from a "walking rate book" to a respected expert involved in higher-order planning and action.[154]

The industry aggressively publicized the narrative that computing machines would liberate knowledge workers from the drudgery of clerical chores and tedious calculations, allowing them to concentrate on more creative labors—a rationale that has been remarkably durable.[155] This talk flattered people in research and media management positions, whose judgment was said to be irreplaceable and whose autonomy and authority would be enhanced by automation. But with agencies almost always looking to save on salaries, these reassurances probably rang hollow to workers who did not have "vice president" or "director" in their titles. Reports suggest that agencies made substantial cuts among "low-skill" staffers, even as they hired personnel trained in the use of computers.[156] Specialized information services also threatened agencies' clerical and tabulating workers. Central Media Bureau, a company that provided administrative and data processing functions in spot buying, reportedly reduced one agency's annual clerical costs by 43 percent.[157] These outside vendors had no loyalty to agency employees; replacing these workers was part of their value proposition. Promoting Simulmatics in 1963, Ithiel de Sola Pool insisted that computers could think faster and remember more than human laborers, without ever needing to eat or sleep or take breaks.[158]

It is ironic that the description of computers as modern conveniences that spared workers from "drudgery" mirrored the appeals made to female homemakers to buy electrical appliances and preprocessed cooking and cleaning products.[159] Although the trade press did not confront the issue, these characterizations of automation and job loss had a gendered dimension. When Y&R and BBDO "unleash[ed] media computerization," *Advertising Age* reported, "both agencies denied rumors that computers will take over media men's jobs. 'Only statistical-type positions will be affected.'"[160] The term "media men" may be an artifact of the period, but it is conspicuous here, since those "statistical-type positions" were often staffed by women. The framing of automation as a transition away from tedious, substitutable labor and toward creative, specialized expertise reflected a masculine and managerial enclosure of the scientific media work that was gaining in professional status.

Both the gendered dynamics of computing in advertising and the resistance to and normalization of automation in media buying and selling are important areas for further research. Throughout advertising's evolution, we see this tension between managerial efforts to eradicate the incalculable human element from economic activity and workers' reluctance to be eradicated from their jobs.

CONCLUSION

"Computer systems have always been sold with the suggestion they represent a ticket to the future," explains Thomas Haigh.[161] The advertising business put that sales pitch on dramatic display. Starting in the 1950s, the materiality and organization of advertising operations were adjusted to accommodate new resources for accelerating and reformatting calculation and exchange. The potential to automate and optimize media buying and selling framed computer technologies both within existing routines and as part of a bright but contested future. Electronic data processing and computer networks were interpreted and leveraged to handle the administrative and informational load in a marketplace where complexity and rising costs motivated the search for analytical and logistical advantages. In addition, computers were used to advance marketing and audience manufacture toward the dream of more rapid and rational action. Advertising took its first clumsy steps toward online marketplaces and machine-to-machine transactions under the supervision of an emergent class of technocratic managers, systems experts, and equipment or information vendors.

In the construction of programmability, computer systems were cast as both an infrastructure and a metaphorical template for information-intensive markets. Computerization furnished an image of audience manufacture as a set of high-speed data processing and trading operations; it also provided the material basis for those arrangements and flows. Computers and the management sciences so closely tied to their adoption in American business provided the technological and discursive tools needed for media planning and buying to rebrand itself as a calculative, data-driven discipline suitable for algorithmic optimization and control. Performances of machine efficiency lent credibility to risky and uncertain investment decisions, and they gave authority to the decision makers who could command the rationalizing force of an information technology revolution.

A calculative culture oriented around automation and optimum seeking fueled changes in media services, the composition of agency personnel, and the power dynamics that ultimately determined which sources of news and entertainment were profitable. "A small army of mathematicians and scientists has captured a share of agencies' and advertisers' media work," one writer observed in 1962. "The 'lush life of three-hour-luncheons-and-cocktails' may be over for some media salesmen, whose statistics are being increasingly scrutinized by computers."[162] The president of a leading rep firm admitted in 1969, "The old basis of buying on personality alone had 90–95% disappeared because of the increased demand for information."[163] Specialist media buying agencies emerged in the mid-1960s, promising campaign costs anywhere from 10 to 40 percent below what media departments at full-service agencies delivered. By 1968, these buying agencies were handling billings estimated at $225 million, with the majority of that spending concentrated in spot TV.[164] Initially, they courted clients by using bulk buying to realize cost efficiencies and by undercutting full-service agencies' 15 percent commission on media spending. Gradually, though, these buyers promised scientific techniques for optimizing expenditures and they increased the pressure within the industry to produce quantitative data that could be fed into computer models.[165] Technical expertise became a hallmark of their role in the value chain and a lever of influence in the media system. Two media agency executives predicted in 1976 that "productivity and prosperity in the 1980's will belong to those who can harness the technology of the computer."[166] Assessing the "era of data explosion" that followed, Helen Katz concluded that "the extent of computerization in media departments," which outpaced computer adoption in other agency departments, "suggests they are in the vanguard of change."[167] By the end of the 1980s, data suppliers and technology vendors were marketing dozens of computer systems to media departments and helping to "make this data explosion more manageable," as a JWT memo put it. The agency compiled a catalog of nearly forty computer systems available for media planning and buying in 1989.[168]

The computerization of media decisions and transactions was a bridge to the future of adtech. Early discussions about automation reflected and informed frameworks for thinking about advances in advertising that became central to the industry, albeit not without contradictions and controversies. Although these aspirations were mostly disappointed in the short term, the discourses analyzed in this chapter shaped the meanings and possibilities surrounding information technology and calculative techniques. The next chapter looks at how marketers and ad sellers continued to chase efficiency, this time by dividing consumer populations into smaller units. Companies selling cable audiences tried to package specific households so that advertisers could exclusively engage the people they deemed valuable. This involved building logistical utilities for the twin purposes of selling audiences and selling sponsors' products.

6 ADDRESSING THE AMERICAN PERSON: DESIGNS FOR PRODUCING AN AUDIENCE OF ONE

In a generous assessment of foresight, business historian Margaret Graham says the invention of the transistor made it "possible to foresee the arrival of truly individualized (or at least customized) information, and with that a resulting fragmentation of the collective experience."[1] This chapter examines some early attempts to make way for that arrival in the business of selling the American people. Continuing to view advertising and audience manufacture as problems of information, distribution, and logistics, I argue that efforts to target individual households via cable television brought a direct-marketing paradigm toward the center of the advertising and media industries and thereby helped build a foundation for the particular mode of commercialization that now dominates the web and much of the mobile app economy.

WASTE MANAGEMENT

An interest in disaggregating mass audiences into lifestyle and behavioral categories has been an explicit element of marketing for more than sixty years.[2] Some of the biggest and brightest lights in the US economy were fixated on this sort of advertising efficiency in the middle of the twentieth century. DuPont, General Motors, and other industrial titans were "trying harder than ever to trim waste from ad budgets and zero in on potential customers," the *Wall Street Journal* reported in 1963. "Many advertisers believe they are increasing the return on their ads by deciding what sort of people are most likely to buy their products and then limiting their campaigns as much as possible to these groups."[3] Expressing the hope that units of audience attention could be packaged for more selective buying, the director of advertising at General Motors said in 1964, "I think the advertiser will be able to buy a pound of people or a pound of territory as today you can buy chicken breasts or legs without getting stuck with the whole bird."[4] Energized by the demand for expert services implied by this "mantra of efficiency," to use Jennifer Karns Alexander's phrase,[5] advertising agencies and marketing consultants sold solutions that appeared to leverage the power of computers and management science.

A community of advertising professionals began to articulate the discriminating affordances of information technology. New tools promised to let marketers and attention merchants organize more specific and finely appraised consumer segments and to isolate them from undesirable consumers. Segmentation and computerization made obvious allies. *Advertising Age* began its coverage of a 1964 seminar with the headline "Computers Open New Vistas in Pinpointing Audiences."[6] Welcoming the trend toward data-driven targeting, one media director observed, "the machines are simply another tool in a continual search to refine information as to who are the clients' best prospects and which are the media that can best speak (and *as exclusively as possible*) to these prospects."[7] Another agency manager said the "key" to extracting efficiencies from his company's computer system was to "delineate product buyer profiles with extreme accuracy."[8] Essentially, proponents claimed that the "precision and control afforded by computers" enabled advertisers and their buying agents to "zero in on audiences that are the most lucrative prospects for their products."[9]

Working with these computer systems, operations researchers designed media optimization models to formalize evaluations of audiences and advertising opportunities. The models were supposed to help media planners and buyers "objectively" discern the expected value of each decision alternative and thus make optimal choices. As media managers often insisted, mathematical modeling and computerization set up conditions for media selection to be reoriented around more detailed profiles of audiences and consumer populations. The goal was to identify increasingly bespoke audience segments and to judge advertising investments based on the success in reaching those specific groups. "The media planner's number one job is getting his messages through to honest-to-goodness prospects, with as few messages to nonprospects as possible," explained a report from the Advertising Research

Foundation.[10] As one observer put it in 1963, "If [the advertiser] wants to reach part-time working housewives, age 35–49, in the $5,000–8,000 bracket living in multiple dwelling units in market No. 6—and the data is available relating this group to media covering market No. 6—the computer will be able to recommend the 'right' media approach." These calculations heralded the emergence of "a highly specialized form of media buying where the rule is cost-per-thousand 'prospects,' not, as in the past, cost-per-thousand people."[11] Proponents claimed that the optimization techniques discussed earlier could set the media buyer's focus securely on efficiency by isolating valuable consumers. "It's not how much mass he buys," explained BBDO's media director, "but how little waste. Linear programming helps show us how to minimize waste."[12]

Looking forward from 1970, an executive at J. Walter Thompson told his colleagues at the agency, "perhaps the greatest changes of all in the decade ahead will take place in the ways and means of singling out those consumers by individual tastes and interests and reaching them by more specialised means."[13] Such specialized means of reaching consumers are the topic of this chapter. Despite all the hopeful speculation about efficient targeting, the ability to define consumers more precisely was not matched by an ability to differentiate one broadcast audience member from another during distribution. The TV advertising business had to face the practical challenge of discriminating among consumer targets in the routing of commercial messages.

In the process of navigating these ambitions and difficulties, certain advertising and media sales professionals began to define cable television as an addressable technology—capable of distributing messages to identifiable households. Addressability is about discrimination and exclusion/inclusion. It is a technical means of delimiting the population of message recipients, sometimes at the level of individuals or individual devices. In other words, an addressable system can specify who has access to a message and who does not, and it can tell the difference between members of those two groups.

Recognizing and acting on difference is what I mean by discrimination here. It is not inherently troublesome. For example, we want the postal service to be addressable so that people don't receive someone else's mail. Critical judgment should hinge on the organizing logic of an addressable system: according to what values and priorities are people excluded or (perhaps predatorily) included? In this case, a sociotechnical system designed for efficient marketing classifies people based on their perceived value as consumers. Given the intersecting and cumulative inequalities in US society, this form of economic ordering tends toward disparate negative impacts on marginalized groups—particularly as addressability spreads beyond cable TV.

Crucially, a workable system of addressable advertising depends on more than the technical capacity to address a transmission to a specific recipient or device. Equally important is the administrative capacity to coordinate market activity around customized or personalized definitions of "audience." That means establishing transactional processes whereby microtargeted audiences are made material, salable, and legitimate as objects of routine planning and action in advertising. To translate the technical affordance of addressability into an implemented institutional capacity required major efforts to operationalize the individual—to make unique individuals (or their personal devices) into something marketing and media businesses could recognize, package, and act on. This turned out to be a daunting challenge. That challenge exposed a central contradiction in the dream of optimizing audience manufacture: ambitions to define consumers in greater detail and to target them more precisely run up against the complexities and costs of incorporating detail and precision into organizational processes.

The history of addressable cable advertising reveals concerted though often frustrated efforts to develop institutions and infrastructures for producing and selling an audience of one. While the specific actors and technologies described here are particular to their times and places, the making of addressable TV advertising surfaced many of the same issues and affordances involved in building internet advertising. The success of that latter project was achieved in part through infrastructure services that effectively made and maintained a market that trades in individualized audiences and capitalizes on the evidence of attention left behind by their online activities.[14] Google and Facebook came to dominate online advertising in part because they built (and bought) logistical utilities for coordinating advertising markets that feature a staggering number of buyers and sellers and relatively detailed definitions of target audiences. Broadly speaking, the history of digital advertising is about assembling a set of technical and administrative systems to manage discrimination-as-a-business on a massive scale and at instantaneous speed. Though humble by comparison to today's world, this is what companies in the spot cable market were trying to accomplish.

SPOT CABLE ADVERTISING

Spot cable ads are delivered by the local cable operators that typically own two or three minutes of commercial time per hour on each of the cable networks they carry. Just like the broadcast spots discussed in the previous chapter, spot cable inventory is sold by both local staffs and sales reps that liaise with regional or national advertisers. Spot cable was a bottleneck for selectively accessing consumer targets with electronic media—for realizing the dream that neighboring households watching the same TV show could be served different ads. Cable networks such as MTV, BET, and Bravo tried to attract narrow audiences who fit the profiles of wealth, race, ethnicity, gender, age, and lifestyle that certain marketers desired. For example, Bravo advertised itself in a trade magazine by putting its logo next to wire cutters, a lock pick, and a crowbar—a lineup of "tools used to break into the most expensive homes."[15] But those networks had no technical means of discriminating between valuable viewers and "waste" within the audiences they attracted. The ability to address a video message to specific viewers resided instead with the cable operators transmitting programs and ads via their wired connections to subscribers' homes. The move from segmentation to household-addressable advertising was therefore accomplished through cable operators and the spot cable market. The dual problems of isolating particular subscribers and marketing those individuals or households to advertisers manifested contradictions between the size and standardization needed to appeal to brand advertisers and their agencies and the granularity and precision touted as cable's advantage. Navigating between mass marketing and direct marketing occasioned a host of logistical dilemmas. Yet, even while the industry spun its wheels, ambitions for addressable advertising intensified as new digital and interactive devices became a wishing well into which various actors cast their hopes for the future.

A MARCH TOWARD ADDRESSABLE ADVERTISING

Addressability and cable advertising developed on parallel tracks that seldom intersected until the late 1980s. Serious discussions about local cable advertising began around 1970, soon after the Federal Communications Commission proposed new rules regarding the development of cable television, including a mandate that systems with at least thirty-five hundred subscribers (roughly three hundred of the twenty-five hundred systems operating at the time) originate programming locally.[16] There were doubts that ad sales could cover program costs, which by one operator's estimate would require selling spots for anywhere from four to twenty times the price of a comparable radio spot. Even so, some observers saw a promising future, and a few firms prepared to act as sales reps to national and regional advertisers.[17] By the middle of the decade, more than six hundred cable systems were originating programming, and these operators reported median advertising sales of $7,500 in 1974, with the most lucrative systems claiming as much as $225,000 in sales.[18]

Like in the early days of web advertising, big brands and their agencies budgeted funds for cable tentatively, in part because, as a vice president at J. Walter Thompson discovered, there was "simply no way to evaluate efficiency."[19] Compton Advertising's director of broadcast programming listed two major reasons for national advertisers' reluctance to invest in cable: "the abject lack of statistical information" and the administrative deficiencies that made cable spots cumbersome to buy. "Like the snake eating its own tail," he warned, "the advertising budgets will not include consideration of cable till cable operators organize themselves better. And the latter will not organize into better 'networked' groups until they can see the big dollars coming their way."[20]

As satellite distribution connected cable operations across the country, the early success of advertiser-supported cable networks both inspired optimism about advertising on local systems and magnified the challenges of building this market. According to a former cable engineer, United Press International (UPI) pioneered the practice of allowing operators to insert local advertisements into commercial time on a nationally delivered channel as an incentive for them to carry UPI's news service. The UPI offering was short-lived, but CNN and others replicated the strategy.[21] By 1980, a leading vendor of the equipment used to insert local ads into cable feeds counted eleven networks offering availabilities to cable operators, including ESPN, BET, and USA Network. Already there was an awareness that the "specialization" of cable programming "allows advertisers to 'zero-in' on a selected target audience."[22] But only about 15 percent of cable systems were generating ad sales, and cable's $60 million in *total* advertising revenue (including network advertising) looked puny next to subscription income exceeding $2.5 billion.[23] At just $8 million, spot advertising constituted about a third of 1 percent of that combined revenue. A sales manager at one of the first cable systems to build a successful advertising business put the incremental revenue he helped deliver into perspective, saying, "It takes my 35 people a whole year to get the same amount of revenue that the system gets in one month from subscription fees."[24] Advertising sales therefore ranked low among cable operators' priorities. Stating the matter frankly, the chief financial officer at Warner Communications told managers from the company's cable division in the early 1980s that their local advertising revenue did not even register as a rounding error on the conglomerate's balance sheet.[25]

Making this market would take some dexterous work (see figure 6.1). Gerry Levin of Time Inc. urged the cable industry to commit to establishing

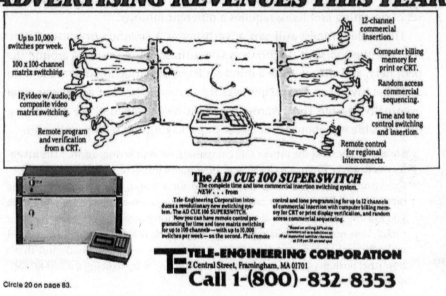

FIGURE 6.1
This advertisement for cable ad insertion equipment gestures toward the complexities of the process and the importance of automation. Though spot cable was still in its infancy, the ad touches on logistical capacities that would become central to more sophisticated ways of serving ads, including remote control of interconnected operations, random-access sequencing of spots, and integration with automatic verification and billing systems. It might be too much of a stretch to see this many-handed machine as an analogy for digital adtech. (*Source: Cable Television Business*, April 15, 1983, 32)

bedrock organizational capacities. Encouraging operators "to build in the infrastructure simply to make advertising sales meaningful at the local level," he nevertheless warned that this was "going to take a few years. So anyone who has entered the business with the notion that there will be a lot of advertising revenues early on its [sic] going to be disappointed."[26] By 1984, more than eight hundred cable systems had "local advertising sales capabilities," and at least thirty technology vendors claimed to "design and build automated commercial insertion equipment for the cable industry."[27] But selling spots to national advertisers remained more challenging than many had hoped. A financial analyst elaborated the "problems and potentials" for cable advertising, which he considered the "most important new cash flow source for the cable operator." One of those problems was the organizational culture: "selling advertising is not in the mainstream of 'traditional' cable operations. It requires a different technology, creates a different set of operating problems, requires a different mindset."[28]

The cable business still saw advertising as a sideshow to subscription. But over the course of the 1980s, a sales infrastructure emerged around the periphery, and with it came a mind-set fixed on extracting new value from available inventory. Larry Zipin was the corporate head of ad sales at what eventually became Time Warner Cable. As he explained it, the maturation of spot cable was about refinements in packaging and selling audiences:

> Me and my peers at the other cable companies, we kept saying, again, "The inventory is finite at the macro level. But, how many different ways can we slice it?" Cuz, you know, if you sell a loaf of bread for a dollar, you get a dollar. If the bread has 50 slices and you can sell each slice for a dime, how much money are you making? Now, if I can slice the bread 50 [more] ways and sell each one for a nickel, how much am I making? . . . We kept slicing that loaf of bread and started making more and more money with the inventory we already had. And that's how [the] business goes from being a mom and pop shop to being a multi-billion-dollar enterprise.[29]

As we will see a bit later, though, slicing that inventory was no piece of cake.

Meanwhile, closer to the core subscription business, cable operators were developing addressable capabilities. Addressability allowed operators to discriminate between households that were authorized to access certain products or services and those that were not. Addressable set-top boxes (STBs), sometimes called terminals or converters, were installed in subscribers' homes and connected to computer control systems at cable operators' headend distribution facilities. These technologies worked together to administer "conditional access," meaning that the central computer could determine whether a specific household's STB was eligible to receive a given signal. "In an addressable system," *Broadcasting* explained, "the headend and the home terminals are tied together by a computer," enabling the operator to control the system remotely.[30] As a 1983 advertisement from Zenith put it, the company's addressable terminal let operators "'talk' to over one million individual subscribers from the headend."[31] Addressability helped operators exclude nonsubscribers from premium channels such as HBO and execute certain pay-per-view functions. It also allowed customer services to be initiated, altered, or discontinued with a few keystrokes at the headend, instead of sending technicians on expensive "truck rolls" to customers' premises.

By the 1980s, home terminals and the computer equipment for headend facilities were becoming more sophisticated and affordable. Operators could expect to pay approximately $135 for an addressable converter and $150,000 for the computer control system.[32] Though still an extravagance for most operators, these costs were within reach for some of the large companies that managed multiple cable systems. Addressability thus began to gain a foothold.[33] According to one estimate, about 500,000 homes were connected to addressable technology in 1983.[34] By 1990, some 9.3 million addressable STBs had been deployed.[35] Homes equipped for addressable advertising via traditional pay TV services numbered almost 65 million in 2019.[36]

Although addressability enabled a cable operator to identify and communicate exclusively with each STB, the idea of sending individually targeted video ads was not close to being accommodated by the capacity of the analog cable plant in the 1980s. It would have required reserving channels exclusively for advertisements, and that made no financial sense when state-of-the-art systems had fifty-four to sixty-six channels, and most operators had thirty-six or fewer. Still, fairly early on, observers glimpsed portents of profound change.

By the early 1970s, the cable industry recognized that home terminal equipment, networked to an operator's central computer, could be the gateway to a suite of futuristic functions, including interactive (two-way) communication, e-commerce, and market research.[37] The general manager of planning for Pioneer Corporation, a maker of cable system equipment, cast addressability as the foundation for a transformative process of media convergence. "At Pioneer," he wrote in 1981, "we view 'Addressable Control' not as an end, but rather the start of a new beginning for cable. . . . A first big step toward the marriage of computer, cable, and the consumer. . . . Could it be that the computer technology called 'Addressable Control' will eventually breed such changes in our industry that some day 'cable television' might even be a misnomer for the services we provide?"[38] Scientific-Atlanta Inc. played on the theme of convergence the next year in a brochure promoting its addressable STB: "Scientific-Atlanta introduces the computer that thinks it's a set-top terminal" (figure 6.2).[39] In describing a packet-switched

FIGURE 6.2
This 1982 promotion for Scientific-Atlanta's 8500 addressable terminal, powered by microprocessors, informed cable operators that its "central billing computer can 'talk' to the addressable 8500" by way of an "intelligent control unit." (*Source:* Barco Library, Cable Center, University of Denver)

network for two-way data communications, Paul Baran, one of the architects of the internet, suggested that cable operators' need for "addressable converter capability to control pay TV delivery" would allow new interactive services to "come along as a byproduct."[40] Although their functionality was limited, addressable cable boxes were among the first networked computing devices—reporting to a remote control center and integrated with a customer management database—introduced into millions of American homes.

Cable operators and technology vendors expected addressability to enable direct marketing, but targeted advertising by national brands was not a priority. That began to change over the course of the 1980s. Spot cable was a modest but rapidly growing business. Cable systems recorded gross advertising billings of $8 million in 1980, $167 million in 1985, and $634 million in 1990. By 1995, billings exceeded $1.4 billion.[41] Spot advertising had become more important to the cable industry as new subscriptions plateaued and it faced rate regulation of its main revenue source in 1992. With "advertising loom[ing] as among the most promising lines of business not touched by government oversight," a cable sales manager from Atlanta said in 1993, "we're going from the ugly stepchild of the cable industry to a group they'll bow in reverence to."[42]

The most successful ad sales operations were generating up to $32 per subscriber per year at the start of the 1990s, and by the mid-1990s, spot advertising was contributing significant free cash flow to company ledgers.[43] At Time Warner Cable, it represented the third most important source of cash flow, behind subscriptions and consumer equipment but ahead of pay-per-view.[44] In 1996 one observer called local advertising "the fastest-growing portion of cable."[45] No less a figure than John Malone, the CEO of Tele-Communications Inc. (TCI), trumpeted its significance to cable systems, projecting that the percentage of TCI's cable TV revenue represented by advertising would triple between 1993 and 1998.[46] TCI reported a 27 percent increase in ad sales in 1994, and it soon hired more staff as "part of a sweeping new emphasis on its advertising sales operations."[47] Industrywide, annual spot revenue quadrupled during the 1990s, finishing the decade at over $2.6 billion.

Interest in addressable advertising increased with the growth of cable audiences, maturation of the spot business, and optimism about the computer processing power of digital STBs. In 1989 General Instrument, which made STBs for TCI and other operators, applied for a patent covering a "method and apparatus for providing demographically targeted television commercials" on cable systems through an addressable converter.[48] In a 1992 report, engineers at Cable Laboratories Inc. (CableLabs), a technology research and development consortium, pointed out that the ability to serve targeted ads gave cable a structural advantage over emergent direct-to-home satellite services: "The cable industry is uniquely positioned to implement an architecture which enables advertisers the ability to 'precision market' its [sic] products on the basis of geographical and demographical boundaries."[49] Bill Harvey, a veteran researcher who had worked on optimization models at large ad agencies, was among those trying to activate this potential. In 1993 he told *Cable Avails* magazine, "As soon as you add interactivity and address-ability you've got a direct response medium. . . . The ability to database and manage this subscriber base will move spot cable beyond traditional TV. . . . Spot cable will become the most interesting medium out there."[50]

Harvey and other enthusiasts appreciated digital STBs' potential to cus-tomize spot cable commercials for specific viewers. "That's the promise of advertising executives," one journalist wrote in 1996, "who say that new digital technology will not only let cable spot advertisers reach geographic zones but will someday create the cable equivalent of direct mail: individ-ualized ads targeted to neighborhoods and households." The media direc-tor at TBWA Chiat/Day warned, however, "It will take the commitment to invest in the technologies that allow us to deliver these ads to an individual household."[51] And the challenges of addressable advertising were not only technological. The spot cable business also needed to establish transactional norms that smoothed the edges between a direct-marketing paradigm and the established habits in mass-media buying. The corporate media director at Hal Riney & Partners distilled national buyers' perspective on new digital advertising systems in 1992: "I think we'll beat a path to their door," he said of agencies' interest in what cable systems might be able to offer, "but they've got to play by the same rules that we use to buy spot television."[52] Next, we look at how the spot cable business negotiated those rules and assembled itself into a billion-dollar sector of audience commodification.

PACKAGING CABLE AUDIENCES

From the mid-1980s onward, cable operators' ad sales departments tried to appeal to brand advertisers while also accentuating cable's direct-marketing capabilities. This led to a familiar contradiction. Audience manufacture called out for standards, protocols, and routines, but the thrust toward personalized direct marketing complicated these transactions and the metrics for evaluating them. Several developments shaped spot cable in ways that provide insights into the broader evolution of targeted advertising.

GETTING TO KNOW THE AUDIENCE

Advertisers and agencies pressured cable operators to make their audiences easier to measure and buy. Advertising agencies generally preferred to transact around the same metrics used in the market for broadcast audiences, although some buyers tolerated a certain amount of flexibility, based on the belief that cable was suited for targeted rather than mass advertising. Even in the latter case, though, advertisers and agencies insisted that the "cable industry must supply more subscriber data," including "thorough statistical profiles" of the people and communities they served.[53] The president of the Cabletelevision Advertising Bureau (CAB), a trade group representing the industry's advertising interests, expected Nielsen ratings to continue to be "the strongest currency in the television marketplace." But, like the champions of optimization models, he saw an opportunity to move from a mass-exposure paradigm toward one with greater sensitivity to business objectives and outcomes: "The challenge is to process and analyze audience data in a more precise relationship to marketing goals."[54]

Broadcasters liked to remind advertisers of the deficiencies in cable audience measurement; yet many people recognized that cable was ultimately in a far superior position to generate data about its viewers.[55] As early as 1982, a letter to the editor of *Broadcasting* insisted that STBs—especially "intelligent" devices equipped with microprocessors—could be used to "poll" the entire cable system and measure precisely who was watching what. "Amazingly," the author wrote, "such on-line, realtime data gathering capability is within reach of the industry using currently available technology."[56] Equipment vendors used this as a selling point. A 1992 catalog for Pioneer's addressable equipment told cable operators, "Viewer Statistics . . . can be used as an advertising and marketing tool which allows you to take five global or individual 'snapshots' of the programming subscribers are viewing at a given instant in time."[57] Already by 1988, Pioneer's marketing materials described these lists of "the programming every subscriber watched at a single instant in time" as "invaluable information for advertisers on subscriber viewing habits."[58] When General Instrument patented a technology for demographic targeting of TV ads in 1992, it noted addressable systems' ability to uniquely identify and observe individual households: "Demographic data can be input by a viewer via a remote control, downloaded to a subscriber's converter from a remote headend, or programmed into the converter at installation. . . . Statistical data can be maintained concerning the number and identity of subscribers viewing specific commercials."[59]

With new forms of user interactivity being imagined and built into cable systems, the flow of consumer data to operators and their advertising partners promised to cascade beyond previous limits. "Every time a viewer presses a button to make a viewing choice," *Broadcasting* reported in an article about an interactive TV venture financed in part by Coca-Cola, "demographic information is recorded and a custom-selected commercial can be shown to that viewer."[60] That venture, called ACTV, received a patent in 1986 for a method of creating user profiles to inform the real-time selection of content delivered over an interactive cable channel, including "tailored television commercials for particular subscribers."[61] Even apart from their interactive services, cable operators were amassing enormous databases as a matter of course. A sales executive boasted in 1988 that the cable industry possessed "a lot of hidden assets." As *Broadcasting* elaborated, "cable operators have gathered extensive information about their subscribers—demographics, likes and dislikes—through their everyday business." These databases, the sales executive said, would enable cable operators to combine "the power of television with the effectiveness of print."[62] Bristling at criticisms of cable audience research in the early 1990s, the president of Cox Cable fired back at an agency executive, "Are you really ready for all the information we have?"[63]

By the end of the 1990s, the continuous passive collection of data via STBs was almost completely normalized. The chief technologist at a firm called Invidi, a leading provider of addressable ad technology to the television industry, explained that the impetus for the company's creation in around 2000 was a desire to use STBs to figure out who was watching and how to advertise to them. "The original concept was: through monitoring behavior you could anonymously determine the likely age, gender, maybe even income and education level of the current viewer and use that information to target ads against. So, early on, we did a tremendous amount of machine learning research to come up with algorithms that actually could give you those kinds of guesses."[64] Invidi and other adtech companies were developing software to document each interaction between users and their STBs. By the mid-2000s, a company called Navic Networks (later bought by Microsoft) had deployed this capability in roughly ten million homes, automatically generating daily data reports about every subscriber's activity. Among other uses, these data were leveraged by spot ad sales staff in negotiations with advertisers and used for demographically targeted interactive advertising.[65] Companies like Nielsen, Rentrak/comScore, and TiVo, as well as operators like Comcast, went on to collect STB data for both audience measurement and consumer targeting. The surveillance capacities of cable have come a long way toward accomplishing what advertisers demanded and operators promised.

Across the 1980s and into the 1990s, we see large cable operators recognizing the trajectory toward personalization and direct marketing, the utility of customer data for advertising purposes, and the advantages of their position as both data collector and gatekeeper of wired connections into individual homes. Given this recognition, it seems ironic that the other major innovations in spot cable ran in the opposite direction: *aggregating* audiences. Three interrelated developments dealt with this issue: clustering, repping, and interconnection. The main thrust of these efforts was to coordinate the spot cable market around advertisers' demands for "one-stop shopping" for segmented audiences. In many ways, these moves prefigure later developments that cemented internet advertising markets.

CLUSTERS, REPS, AND INTERCONNECTS

For cable, geographic specificity was both its value proposition and its limitation. The pitch to a local grocer was simple: why pay to broadcast your ad to all of Philadelphia (and a chunk of New Jersey) when you can use cable to target the suburb where your customers live? For national marketers, though, buying ads on local cable seemed expensive on a per-viewer basis, and reaching beyond the small territories covered by each cable system meant dealing with a swarm of salespeople and a mess of incompatible traffic, billing, and insertion systems. Observers complained of "administrative headaches" and "logistical nightmares."[66] Reflecting on the situation in the 1980s, one cable ad sales expert told me, "If you said, 'OK, I want the first spot in *Sportscenter* on the eighty cable systems in Los Angeles,' and they all agreed to reserve it for you, by the time you did the eighty negotiations, executed the eighty contracts, distributed eighty sets of content, reconciled eighty invoices, and wrote eighty checks, it cost you more than buying one

spot on the CBS evening news. Because the execution costs were astronomical."[67] DDB's vice president of video technology and programming said in
1983, "Local cable advertising probably isn't going to be a big business until
you can buy local cable in virtually all cable systems across the country easily."[68] If cable operators failed to interconnect their systems and assemble
sales networks, one observer warned, their business would be "stuck with
mom and pop shops and nickel and dime spots."[69]

Soon, however, the typical ownership structure in the cable industry
shifted decisively from "mom and pop" companies toward the multiple system operators (MSOs) that still dominate today. MSOs sought the advantages of "clustered" operations—owning adjacent systems across a city or
region. Clustering was mainly about centralizing management and customer service, but it also benefited ad sales. By the early 1980s, cable giants,
such as Time Inc.'s ATC and Westinghouse's Group W Cable, recognized
that clustering would allow an MSO to become a "legitimate competitor"
to newspapers, radio, and broadcast TV in the market for local advertising.[70]

Those MSOs partnered with sales rep firms in the second half of the
1980s, expanding their ability to transact with national and regional
advertisers and their agencies. Spot cable reps had emerged soon after local
operators began inserting ads. In 1983 Cable Networks Inc. (CNI) generated
about $4.5 million in ad sales in the New York area, and four years later it
was bought by Cablevision (now Altice). CNI's chief rival, National Cable
Advertising (NCA), posted just over $7 million in revenue in 1987, and the
next year three MSOs took an equity stake in the firm.[71] NCA merged with
Katz's Cable Media Corporation in 1995 to form National Cable Communications (NCC), which still represents cable operators to national advertisers,
although its owners (Comcast, Charter, and Cox) rebranded it as Ampersand
in 2019. By the end of 1996, NCC represented roughly 60 percent of cable
systems in the United States.[72] When TCI bought into NCC in 1998, the latter's marketing footprint covered ninety-seven of the top hundred markets
and included 82 percent of the cable households receiving locally inserted
ads.[73] By then, NCC and CNI handled about $350 million in total billings.[74]

Sales reps helped cable operators mimic the coverage area and buying
protocols of broadcast stations. But MSOs and rep firms also demonstrated
"an awareness about upgrading technology to develop more psycho-
demographic information." Trumpeting cable's direct-marketing ambitions,
the president of CNI declared in 1989, "It's no longer cost per thousand. It's

cost per ZIP code."[75] National cable reps were an organizational embodiment of the bidirectional transitions and translations between mass and
targeted advertising.

A related development was the interconnection of cable systems, including
those operated by different MSOs. Interconnection expanded coverage and
consolidated ad sales operations. This seriously reduced the complexity and
transaction costs of spot cable. An advertiser could now buy spots across all
the cable systems in a city or region through one point of contact. Interconnections were described as being either "soft" or "hard." In soft interconnections, tapes containing advertisements would be physically circulated (or
"bicycled") from system to system for simultaneous insertion into a network
like ESPN. More significant were the hard interconnections, wherein systems
were networked using microwave relays, shared satellite links, or coaxial and
fiber-optic cables. These arrangements were managed by specialized companies called, obviously enough, interconnects. Interconnect companies provided a legal expedient for cable operators to compete collectively against
local broadcasters without violating laws prohibiting collusive price fixing.
The legal argument was that an interconnect marketed a product that local
operators could not produce and sell on their own—a potential audience of
all cable households within the network.[76]

Perhaps the most innovative interconnect was Adlink, which was
formed in 1988 and eventually connected more than seventy-five cable
systems in Los Angeles and Southern California. It was owned by five cable
operators and a regional sports service that had a transponder on the Sat-
com IV satellite, which Adlink used to distribute ads to its interconnected
systems.[77] Adlink reported revenue of almost $30 million in 1995. Though
this was a tidy sum, as one report noted, it was "only the proverbial crumbs
that fall off the table" in Los Angeles's $1.2 billion TV ad market.[78] Adlink
had built itself to be like "the ninth TV station in Los Angeles,"[79] but by
this time, it was beginning to understand local cable as competing for the
nearly $5 billion spent on *all* media in the city, including newspapers and
direct mail. Its ambition was rewarded with surging revenue—$62 million
in 1997 and $83 million in 1998.[80] This strategy illustrates two moves: scal-
ing up a mass sales infrastructure and then refocusing on direct marketing.
We will revisit Adlink's addressable efforts after discussing how interconnects
adopted digital adtech.

DIGITAL TECH BEFORE THE DIGITAL TRANSITION

There is no denying the importance of digital technology to targeted advertising. Bruce Anderson described the realization he and others had in the early and mid-1990s, when he was the managing director of digital television at Sarnoff Corporation:

> Once you packetize the video stream, it's pretty obvious you don't necessarily need to send the same information to every receiver. So, even back then, there were people starting to think about how you possibly do addressable or targeted advertising. The TV industry, pretty much since it started having paid advertising, was looking for ways to have specific audience segments see particular ads. . . . Ultimately, what you'd love to do is be able to have a one-to-one conversation. So, DMAs [designated market areas] became chunks of cities, which became neighborhoods, which now, with the advent of true addressable advertising, is down to individual households. . . . The switch to digital is really what accelerated the whole process.[81]

The transition to digital television transmission was slow.[82] But a decade before programming became digital, cable operators integrated digital technologies into their advertising infrastructures. Digitization had significant implications for two key logistical functions: ad insertion and trafficking.

In 1992 representatives from the CableLabs consortium explained that "new technology platforms based on integrating compressed digital video mass storage systems and powerful communication network architectures should allow an infrastructure which supports future revenue growth for cable advertising sales."[83] The authors understood the importance of coordinating flows of information and commerce in the spot cable market, and they pointed toward the sorts of services that later emerged to facilitate online advertising: "the purpose of creating this [digital commercial insertion] network is to promote the *interconnection and interoperability of hardware and software* which subsequently allows advertisers to buy local, regional, and national spot avails easily and conveniently."[84] They added, "the convergence of cost-effective computing plus the use of cost-effective communication networks" could provide the "enabling architectures" for solving difficulties in trafficking, inserting, and verifying spots.[85]

With experts convinced that "multichannel, digital ad insertion [was] on the horizon," industry groups began to mobilize. Engineers solicited plans for "a national highway and reservation system" that would route ads to headends around the country. They also sought proposals for hardware "to store and access thousands of spots for insertion" into more and more channels.[86] A technical director at CableLabs estimated that cable operators could boost their annual spot advertising revenue from $1.50 per subscriber to as much as $25 per subscriber by replacing videotape equipment with digital storage and insertion systems.[87] Analog insertion generally required more work, and it afforded less capacity in the number of spots available and less flexibility in how spot inventory could be sold to buyers. As Paul Woidke of Adlink explained in 1993, "In a tape-based insertion environment today, you require three to four tape decks for every insertion channel on which you want to do full random insertion—that is to say be able to hit every avail with a different spot."[88] The number of cable networks was exploding in the early 1990s, and managing "as few as four 30-second spots across even 30 channels (120 total spots) per hour" was considered "cost- and space-prohibitive" with videotape equipment.[89] A central server linked to affiliated systems could take over the "command and control" function of routing ads into content streams. With this "digital revolution" in cable adtech, interconnects looked to provide "a more efficient, and a more effective, and a more economical means of spot delivery in the cable universe."[90]

By the mid-1990s, the largest interconnects were "adopting systemwide server-based insertion systems."[91] They paid anywhere from $1 million to more than $10 million to install and integrate digital equipment across their facilities. Sony and Channelmatic, a leading tech vendor, provided the Chicago interconnect with video serving and switching systems capable of storing twenty-five hundred thirty-second spots, inserting on eighty channels, and splitting the interconnect's territory into five zones for targeting.[92] Although cable systems did not yet transmit digital video to customers (the digital ad files had to be converted back into an analog signal at the headend), this back-end digital infrastructure provided greater storage capacity, more flexibility in routing ads to affiliates and retrieving them for insertion, easier editing and splicing, and a more comprehensive and automated system for verifying that ads ran as promised. Centralized digital video servers also helped cable operators reduce the turnaround time needed to incorporate new ads into the insertion lineup—going from days to hours. As *Broadcasting & Cable* reported in 1996, "digitalization of interconnects, such as Adlink in Los Angeles, has removed many of the barriers and disadvantages that dissuade advertisers from choosing cable over broadcast."[93]

These advertising systems were part of larger developments in server-based storage and retrieval of video assets. Most famously, Time Warner spent a fortune on an ambitious project called Full Service Network (FSN), which deployed futuristic multimedia applications in four thousand homes in Orlando, Florida, in 1994. Excessive hype and runaway costs blunted any perception of success, even though Time Warner did actualize video-on-demand. The company coupled digital compression and storage technologies with the installation of fiber-optic cable infrastructure. According to Jim Chiddix, Time Warner's top technologist at the time, "we had enough channel capacity in each neighborhood where the fiber went to deliver a different stream to each home. And that was really a revolutionary idea." He described the scope of the company's aspirations:

> Time Warner was interested in other ways to use this same technology. This is before the web. The internet existed primarily for email between college professors. And so, we had a vision of using this kind of interactive television for things beyond movies-on-demand. And so, in this trial we launched in Orlando, we had interactive applications. . . . And one of the things we wanted to do was to deliver, to insert different commercials into the feeds for different homes. It was just sort of a germ of an idea; but the same servers that we were using to deliver video-on-demand could also insert commercials-on-demand. And we didn't take that idea very far, but we did play with it. We had a demo that showed it.

"From a technology standpoint," Chiddix explained, "video-on-demand and targeted ad insertion are almost indistinguishable. You're really doing the same thing. You're switching a stream from a server to a given neighborhood and an individual set-top box in the home." The difficulty was "building a business around those capabilities."[94]

Cable interconnects took steps in this direction. Adlink led the pack, explicitly imitating direct-mail marketing. Around 1995 it introduced two products—Adtag and Adcopy—that allowed advertisers to tailor messages for geographically (and therefore demographically) defined audiences within the Adlink footprint, then covering about fifty-seven cable headends and 2.3 million subscribers.[95] With Adtag, a bit of text was appended to an advertisement displaying some geographically tailored information, such as the location of an automaker's nearest dealership. Everyone in the Adlink footprint saw the ad, but the textual tag differed across neighborhoods. In contrast, Adcopy allowed advertisers to buy a spot across the whole interconnect but then distribute different versions of the ad through particular portions of Adlink's network. As one writer explained, "Chevrolet might run an SUV commercial on cable systems in L.A.'s beach communities and a spot for its Malibu model in less-upscale areas."[96] Adlink described its service in 1998 as "the innovative approach to television advertising that will help you reach more of the audience you really care about—the customers most likely to buy your products. . . . It's as close to direct marketing as the medium and the technology allow."[97] Adlink's director of marketing echoed the message that this service involved "more than just local cable. . . . It's the closest thing to addressable advertising being offered today."[98] "Our goal," she said, "is to offer advertisers a technology platform, a research process and marketing applications."[99]

Adlink's products were a crude form of addressability. Adtag and Adcopy worked on the principle of "zoning." A market with dozens of headends that had been aggregated for ad sales through the interconnect would then be disaggregated into regions classifiable by income, lifestyle, and other categories corresponding to spatial segregation. Zoning let advertisers "send different commercials to different neighborhoods on the same channel at the same time, allowing them to customize their pitch to fit the demographic characteristics of audience segments within the same market."[100] In effect, zoning made good on cable's promise of geodemographic targeting while working within sales and distribution infrastructures that streamlined buying and insertion processes. But zoned ads did not discriminate among households within the footprint of a headend; they could not target a specific STB. Moreover, an advertiser had to buy a spot across the entire interconnect and then customize messages for different zones. Still, this "was the rudimentary first step toward addressability."[101]

This progress soon aggravated another logistical weakness. As one reporter noted, "Digital insertion makes possible all sorts of initiatives, such as zoning and same-day insertion. But traffic and billing systems have to be able to keep up with new demands."[102] "Traffic and billing" (T/B) refers to the process of scheduling ad insertions, verifying that the ads ran, and managing payment—basically, routing ads, money, and paperwork. The more granular the targeting, the more complex the T/B processes.

"Complexity" was a scary word for a cable industry that started with minimal competence in the logistics of ad delivery. Larry Zipin recalls that in the early days of local cable, when an operator was inserting ads on just a handful of national cable networks, T/B systems often consisted of "two ladies in their thirties or forties, a big white board, and erasable pens."[103] As the business grew, these systems needed to handle more volume and intricacy. Adding geographic zoning and customized ad delivery could multiply tenfold the number of avails managed by a cable operator.[104] One vendor of T/B technologies tried to reckon the administrative challenge: "If you're running 30 networks, 30 zones at an average of 50 30-second spots a day on that, that's 30x30x50 for one day. That turns out to be tens of thousands of spots that you're running in one day."[105] The general manager of advertising at a Time Warner Cable system in San Antonio, Texas, explained that his operation inserted on forty cable channels, selling more than three million spots a year. By contrast, he estimated that a typical broadcast station would run roughly 120,000 spots. "It's not unusual for somebody to buy 5,000 spots a month from us."[106]

In the 1980s cable ad sales managers struggled to convince top executives to invest in the insertion equipment needed to grow the business. By the 1990s, having achieved greater insertion capacity, they needed new control systems to cope with that growth. Zipin's recollections add color to the intensity of the work:

> Keep in mind, if you're a local ABC station, your traffic system doesn't have to be very complicated because you're only trafficking on one station. And you're trafficking, you know, two minutes an hour, for twenty-four hours a day, for seven days a week. . . . That's logical, sequential—in a straight line. Ours is four-dimensional. And then it becomes twelve-dimensional. And then it becomes twenty-dimensional. So . . . the cable company gets generous and gives us enough ad insertion equipment to go on twenty networks, and [now] I have two ladies in the traffic department who are ready to commit suicide. We literally have total chaos.[107]

Sales reps urged MSOs to "accelerate upgrades of their advertising distribution capabilities, including [electronic data interchange], to allow advertisers to take full advantage of evolving technologies and targeting."[108] Zipin recognized that refinement of the logistical back end was essential for achieving breakthroughs in precise targeting. "What's keeping us from being able to go to specific neighborhood nodes and homes right now," Zipin observed, "is limitations in the trafficking, schedule and inventory software."[109] One report even speculated that new T/B software would elevate the weary traffic personnel from "pure data processors" to "knowledge workers."[110] Given the gendered dynamics of automation discussed earlier, however, it may not be surprising that the women working at whiteboards enjoyed less status than the programmers at adtech companies like DoubleClick who automated T/B for online advertising.

The technical aspects of trafficking were generally understood in the 1990s, and they could be dealt with if organizations committed to the job. What was harder to settle for cable was a conflict about the business model, which had to be resolved before anyone would invest in better software. Addressability revealed competing philosophies about how audiences should be sliced and priced. There were two approaches to slicing. In one model, an avail is sold to a single advertiser that then customizes ad copy for different parts of the population. This is what Adlink was doing. In the other model, instead of selling an avail to one advertiser, any number of advertisers can buy "impressions" within that avail, each representing households that fit certain parameters. The latter version—the eventual model for online advertising—was less popular initially. Its logistics were exponentially harder. Operators worried that even if they could break avails up into targeted impressions and sell them on an individual basis, they risked being left with "remnant" or unsold inventory. In other words, they feared advertisers would buy access to only the most desirable homes. Software to automate the management of inventory could help with the problem, but operators worried that this would drive down prices, especially insofar as it pointed toward using auction mechanisms.[111] Since avails almost always sold out, operators were reluctant to do anything except try to negotiate higher prices.

Battles over pricing also delayed addressable cable advertising. The issue hinged on a dispute about waste. Sellers argued that if they distinguished qualified consumers from the rest of the audience, advertisers should pay more for each viewer. Advertisers and their agencies countered that they were already paying an inflated price because of those wasted exposures, so they should continue to pay the same rates for the people they really wanted to reach in the first place. "The two parties can't agree whether there should be a premium or a discount," Chet Kanojia explained. "That's how far apart they were. . . . The buyer wants a discount because he wants to get rid of excess audience, and the seller wants a premium because he or she is enabling them to do targeting."[112]

The technical details of insertion, trafficking, and pricing could fill another volume. The point here is that spot cable advertising, especially as it approached addressability, gave rise to severe logistical challenges, some of which resurfaced in constructing the internet economy around personalized advertising. The web was unproved as an advertising medium, but it was limited by fewer technical barriers and business incumbencies than cable. Cable operators could only insert addressable ads into the roughly two minutes of spot availabilities networks gave them, and the infrastructure for distributing ads (including the STBs in customers' homes) varied widely in age, capability, and compatibility across or even within individual cable systems. Still, while the web was being groomed for commercialization, the cable industry pushed harder than ever to leverage its legitimizing work and stake claims on the future.

SELLING THE FUTURE

The adoption of digital video servers for advertising delivery ushered in a distinctive period in the development of addressable and interactive advertising. Marrying television advertising with direct marketing became a pervasive ambition, and efforts intensified to manifest it by building software, hardware, and business relationships.[113] Speculation about digital STBs poured gasoline onto hot-burning hype about the convergence of television and personal computers.

By the second half of the 1990s, addressable advertising had entered a new age of action and experimentation. Firms such as ACTV, OpenTV, and Wink implemented new advertising and e-commerce products and helped create a community of apostles who carried their commitments to major positions in media buying and selling. Around 1998 TCI, the largest cable operator in the country, invited a group of blue-chip marketers to its offices in Denver, Colorado, where it proposed that each of those companies give more than $2 million to finance the development of addressable adtech in return for local cable inventory. This summit was the latest in a series of forward-looking performances. Already by 1994, TCI was "promoting the concept of 'cream zones' that will enable advertisers to reach only the households they want: single-parent families, Cadillac owners, frequent fliers or any other consumer profile." Its corporate director of ad sales boasted, "The universe will be as few as one and as many as an advertiser would like."[114] TCI hired a vice president of national ad sales in 1996 and sent him to Madison Avenue. His twofold mission was to proselytize the interactive and addressable vision that John Malone and TCI had been articulating for years and to solicit advertisers' guidance about how such a system could be built to meet their needs. Eventually, this executive came to realize that most of the things he was promising, including the digital STBs designed for the job, existed only as blueprints—what critics today might call vaporware. Years later he concluded, "What I was selling was the future, against your competitor." Most of the advertisers invited to Denver were not much interested in local cable avails. "But they *were* interested in the future of addressability and the future of interactivity."[115]

The future arrived very slowly, interrupted in part by AT&T's blockbuster acquisition of TCI. (Arguably, Wall Street was the most important audience for TCI's futuristic promises.[116]) Despite being hyped as the bridge to widespread personalization, the earliest digital STBs were underwhelming. They had limited memory and the capacity to accommodate only a few crude binary categories for classifying the box and its associated household into marketing segments (e.g., apartment vs. house, kids vs. no kids, income above or below a defined threshold). Adtech software was competing for space on the device against engineering priorities that facilitated cable operators' core subscription business, such as the program guide. General Instrument (GI)—the box maker enlisted to realize TCI's vision—programmed its devices around an operating system that was mostly inhospitable to app developers. GI also subjected potential software and middleware installations to a lengthy certification process.[117] The enterprise of building and deploying boxes capable of facilitating household addressability was therefore a far bigger challenge than boosters admitted. Although it was technically possible, it took a backseat to other priorities, and cable sales personnel still ranked well below subscription-focused engineers and managers.

After several delays, AT&T Broadband (formerly TCI) launched the first major trial of household addressable advertising in 2001.[118] The results were promising, but the protracted rollout of digital STBs denied operators and advertisers the scale they needed for a viable business. That remained the prevailing condition for a decade or so. In the interim, video-on-demand offered a new opportunity to try addressable delivery. Elements of the technology were conducive to targeted and dynamically inserted advertisements, and on-demand access had the potential to annihilate the scarcity of ad spots available to cable operators. But negotiating the rights among operators and programmers—as well as questions about how to reconcile video-on-demand impressions with linear TV's audience ratings—presented impediments.

By the late 2000s, however, a variety of companies—Navic, Visible World, Invidi, BlackArrow, and others—had developed methods of using the data and software in STBs to automate the trafficking and decision-making processes needed to identify and serve a targeted ad to a specific household.

And as the cable plant became digital, operators now able to fit ten video streams into the 6 MHz channel previously occupied by just one analog signal could realistically reserve bandwidth for addressable ads. The formation in 2008 of Canoe Ventures, a well-funded consortium dedicated to this vision, put wind in addressable advertising's sails. When that group disbanded a few years later, other initiatives tried to rally the industry around standards for profiling, pricing, and measuring addressable audiences—and to convince national programming networks to give MSOs more inventory for addressable insertion. Addressable advertising on cable and satellite systems now generates about $3 billion in annual revenue, and the market is much larger if we include internet streaming services such as Hulu and YouTube TV, which have actualized cable's addressable plans. Addressing ads to individual web browsers, mobile devices, and social media accounts is the dominant online paradigm. Household addressable TV advertising muddles on, still "the next big thing," as it has been for thirty years.

CONCLUSION

The internet advertising market sprinted ahead of addressable television, but the spot cable business occasioned developments in a variety of logistical functions that prefigured some basic processes of buying, inserting, and tracking ads in online and mobile media. In overcoming obstacles and chasing opportunities on the horizon, spot cable was a petri dish for the evolution of targeted advertising. That evolution depended not only on the technical capacity to address transmissions to specific recipients but equally on the administrative capacity to coordinate market activity around a personalized definition of "audience." This history shows the importance of establishing interconnections and logistical utilities so that advertisers can buy narrow or customized audience segments without being overwhelmed by transaction costs and complexities. Strategic acquisitions (e.g., Google's purchase of DoubleClick) and the integration of ad serving, avail supply chains, and auction mechanisms across sites and apps suggest that the lesson was applied in scaling up internet adtech. Another key innovation of internet advertising was the automation of trafficking and billing. The online ad business blossomed around databases, algorithms, and identification and tracking technologies for making rapid decisions about ad placements, routing messages to specific recipients, and observing and managing this traffic. On a smaller scale, this is what sales organizations and technology vendors in spot cable were trying to accomplish. They helped usher in an era of personalization by assembling infrastructures and institutions for packaging more finely segmented audiences and by cementing imaginaries about a one-to-one marketing future.

Despite delays and limitations, spot cable's pursuit of addressability planted direct marketing in the bedrock of electronic media and digital advertising. Addressable cable only lunged partway toward making an audience of one; advertisers targeted relatively coarse consumer populations via devices that represented whole households. But it was a move toward making audience commodities more exclusionary along a range of socioeconomic and lifestyle dimensions. Given the intersections of inequality in the United States, the identification of apparently capable and willing shoppers also mapped differences in race, age, gender, education, and class. These differences were operationalized by geography, reflecting the spatial segregation of wealth, status, and power in America.

The construction of addressability was complemented by a related and even more direct strategy for turning audiences into consumers: building a marketplace around them. Addressable targeting was one pillar of a direct-marketing paradigm in new media advertising. Another was creating a return path and transactional apparatus for enabling consumer response. We turn now to the evergreen affordance of shoppability.

7 BUY-BUTTON FANTASIES: THE PERSISTENCE OF SHOPPABILITY

Consumers view a TV commercial and then simply press a button to buy a product—this has been one of the oldest and most widely touted features of interactive advertising.

—George Winslow

New technologies provide a fractal surface where various actors project futuristic visions and try to legitimate their power. Expectations about the economic and social potential of interactive media have been imprinted with particular force onto the policies, practices, and apparatus that make up information and communication systems in the United States.[1] Following Patrick Parsons's invitation to examine the "hopes people have for technology and how those hopes drive business decisions and public policy,"[2] this chapter details the persistence of one prominent theme in discussions about new media futures: shoppability. "CBS founder William Paley once said that television was the ideal selling medium," Robert McChesney and colleagues point out. "Left to Madison Avenue," they add, "the interactive digital world will be the ideal medium for closing the deal altogether."[3] The construction of shoppability in discourses around cable television, telecommunications, computer electronics, and direct marketing reflected long-standing aspirations that continue to shape digital environments. This is a story about how blue-sky dreams became blueprint designs—and how those blueprints were used to powerful effect, regardless of whether the designs were actually built.

SHOPPABILITY AS AFFORDANCE

"Shoppability" is a recent term of art, but its basic premise—that items featured in advertisements and entertainment should be available for purchase—has surfaced repeatedly across the last half century. The idea is to produce consumers by building a marketplace around them, connecting personal and domestic media devices to always-on commercial environments capable of simultaneously stimulating and capitalizing on consumer desires. Pressures to tighten the link between advertising and sales, and to orient attention toward the marketing applications of information technology, have motivated many attempts to engineer interactive and, specifically, *transactive* television systems.[4] These systems are designed to let viewers use their remote controls to buy the things they see in video content, while media companies and market intermediaries share in the revenue or collect service fees. Shoppability has now become ubiquitous on the internet. This chapter historicizes that development by examining efforts to imagine and implement transactivity via cable, satellite, and telecom TV services. It focuses on the integration of *marketing communication* and *marketplace infrastructure* in a single user touch point, illustrating the continuities and tensions that extend from modern influencer marketing back through the promise of "selling Jennifer Aniston's sweater"—a remarkably durable slogan for envisioning shoppable media.

Shoppable television is a glamorous mess. It's a tinsel-trimmed dream that's hit with a spotlight whenever a new video-related technology debuts. As a practical achievement, however, shoppable TV has been plagued by technical hiccups, prohibitive costs, and reluctant consumers. Because of differences in technologies, cultural habits, and institutional and infrastructural legacies, web browsers and mobile apps have been far more successful than television as venues for shoppable entertainment and advertising. Today, ads across YouTube, Spotify, Instagram, TikTok, and many more sites and apps enable e-commerce, as do plenty of the content and interface features that surround advertisements. Social media companies increasingly position themselves as marketplace platforms where sponsors can hire content creators to integrate brands and products into authentic lifestyle performances and where followers can browse and buy from those displays.[5] To understand these entanglements of entertainment and e-commerce, we should take a journey through some moments when old media were new.[6]

Shoppability is an "imagined affordance,"[7] a commercial potential that marketing and media professionals have tried to activate in the convergence of television, computing, and telecommunications. It has been featured in countless stories about the exciting possibilities enabled by technological change. Those stories are revised to meet strategic challenges and opportunities, but their iterations cohere around relatively stable visions of a transactive media future, lending an appearance of inevitability to contingent and contestable developments. From this perspective, shoppability belongs to the history of cable as a new medium. When technologies are new or unsettled in meaning and usage, expectations about their trajectories can have profound, if unpredictable, consequences.[8] Hopeful stakeholders leverage assets and advantages toward establishing legitimacy for visions of development that favor their interests and competencies, trying to harness those visions to economic, political, legal, and cultural power.[9] Tarleton Gillespie notes that firms and industries "frame their services and technologies" in ways that help them pursue business objectives, secure regulatory privileges and protections, and "lay out a cultural imaginary within which their service makes sense."[10] As industries that are capital intensive, structured by government policy, and articulated to financial speculation, cable and telecommunications are acutely preoccupied with the future of technology. Excitement about the advertising innovations of an information age intensified during the development of cable television, reflecting cable operators' competitive advantage of controlling a high-capacity, two-way connection into individual homes. Shoppable television became an aspirational waypoint that oriented ambitions and expectations about media convergence. It directed attention, energy, imagination, and capital toward some (and not other) possibilities for digital communications systems. The following sections identify early examples of this marketing logic and then trace its history in the cable business.

INDUSTRIAL LOGIC AND LORE

AN INTERACTIVE STOREFRONT

Marketing has figured prominently among the design values influencing the technologies and cultural forms of television in the United States. As Jonathan Gray puts it, "a commercial television industry is guided first and foremost by the desire to sell all manner of consumer goods and services."[11] In the 1970s Raymond Williams recognized that within this commercial model, interactivity would tend to be exploited for its marketing affordances. He worried that despite technical capacities to amplify civic participation, two-way TV would confront "reactive consumers" with privatized opportunities, such as "choosing an item from a shop display or from an advertisement."[12] This turned out to be a pleasing and resilient prospect for some influential stakeholders.

In his best-selling book *The Road Ahead*, Bill Gates envisioned TV as an interactive catalog from which viewers could buy anything that appeared onscreen.[13] By 1997, computer programmers at MIT's Media Lab—one of whom was later put in charge of "direct-to-consumer retailing via iTV" at NBCUniversal—had designed a prototype to demonstrate that sort of interactive shopping. They called it HyperSoap.[14] As Gates captured the imaginations of technologists and entrepreneurs, an analyst at an influential consultancy curried favor with marketers. In 2000 Josh Bernoff of Forrester Research predicted that television would embrace HyperSoap-style platforms "in which viewers can buy every item the actors are wearing or using."[15] HyperSoap's name and format seemed to betray a gendered assumption that soap operas and other scripted entertainment would be felicitous venues for selling jewelry and apparel to female viewers who were emotionally invested in characters and story lines or who wished to own artifacts that connected them to admired celebrities. Attaching a desirable personality to the idea, Bernoff planted a seed that continues to attract fertilizer: viewers could buy Jennifer Aniston's sweater.[16]

Selling the *Friends* star's sweater became a trope in industry parlance, setting a hopeful benchmark for interactive TV. HyperSoap was even described in retrospect as having begun "with dreams of buying the sweater off Jennifer Aniston's back."[17] Trade writers repeatedly invoked it as a threshold marking the dawn of entertainment-based e-commerce, and the idea is still summoned—sometimes pejoratively—at industry events.[18] While the plan to sell Jennifer Aniston's sweater disappointed expectations and, in some ways, became an icon for the failed promise of interactive TV, many marketing professionals remain possessed of the idea that television, and most other media, should be a storefront for selling merchandise featured in programming or ads. This extends an almost existential ambition for commercial media systems. As one observer puts it, "The idea of being able to cash in viewer demand for, say, the sweater Jennifer Aniston was wearing in *Friends* has been around longer than *Friends*."[19]

THE DEEP ROOTS OF SHOPPABILITY

Some Americans began to use electrical media for remote shopping in the nineteenth century.[20] Placing delivery orders by telephone, affluent consumers or their servants could conveniently summon groceries, drugs, and dry goods from the comfort of home, sparing themselves the displeasure of encountering racialized strangers or inferior social classes in public spaces.[21] Radio also brought the commercial world into households. American broadcasting has had marketing in its bloodstream since the 1920s, with efforts to manage consumer demand institutionalized in its economic and cultural forms. Recognizing radio's potential as a "major selling medium," national advertisers made some of the earliest sponsored programs to introduce homemakers to branded domestic products and methods for using them.[22] Department stores established radio stations and orchestrated programming around merchandising priorities, sometimes broadcasting from the shop floor.[23] These urban mass merchants had already contributed to what William Leach calls the "democratization of desire," staging spectacular and intimate displays of abundance and leisure.[24] Radio helped them spread the imagery of consumer capitalism beyond their marble walls. One retailer reported that "listeners at home come in to see the things of which [the host] Enid Bur has spoken, with the desire to buy already created."[25]

As early as 1944, Macy's brought this concept to television, scheduling *Tele-Shopping with Martha Manning* (later renamed *Macy's Teleshopping*) on DuMont's WABD New York.[26] By 1949, WTAG Worcester described its product integration strategy as "Sell-a-Vision," bragging that one sponsor "sold out its supply [of scented wrapping paper] in a matter of hours" after a broadcast.[27] Following the template that radio set for daytime programming, television's department store and "magazine" formats addressed themselves to "Mrs. Daytime Consumer." They not only showcased fashionable pleasures but also presented products in ways that encouraged the presumed female audience to treat viewing as part of the work of making "consumer choices for their families."[28]

These ventures all implied that new media should be integrated into product marketing and a mode of social reproduction centered around the efficient management of private households. Broadcasting offered dramatic and seemingly personal ways to animate commodities and consumption, packaging sales efforts and consumer socialization within popular amusements. But unlike the telephone, broadcasting provided no mechanism for initiating purchases. The construction of cable television accelerated ambitions to make the medium itself into a marketplace.

Community antenna TV (CATV) operations began distributing television signals by cable in the late 1940s. Within two decades, entrepreneurs, analysts, and policymakers recognized the prospect of delivering multiple services over an integrated wire infrastructure.[29] Enthusiasts predicted sweeping changes in the production and consumption of information and entertainment, culminating in a "wired nation" where cable would accommodate news delivery, telephony, home shopping, and more.[30] These visions could be seasoned with democratic flavors when interactive technology was interpreted as facilitating wider participation in public affairs or extending access to goods and services among people who were homebound, living with disabilities, or mistreated in retail settings because of their race, gender, appearance, or language. Certainly, the technological material was flexible enough to accommodate these values. But since affordances take shape within particular social relations and priorities, the construction of shoppability reflected the facts that the growth market for cable was upscale subscribers and the electronic media Americans used in their homes were thoroughly commercial. A 1966 article in *U.S. News and World Report* anticipated Williams's assessment of interactive TV, speculating that "merchants will use extra channels to display their wares more fully than they can on the usual spot commercial." A "housewife," the article noted, can "select a dress from the television screen, electronically place her order for the dress, and direct her bank to make the payment."[31]

Interactivity was thus often defined as transactivity, draped in familiar middle- and upper-class consumer imagery. This vision combined the democratization of desire with an impulse toward privatization and the avoidance of embodied social mixing. Whether via online shopping or the apps people use to hire on-demand chauffeurs and personal shoppers, electronic commerce continues to promise a convenient way around awkward human interactions in marketplaces or at the edges of domestic spaces.[32]

Glimpses of this seductive future—like the past, only better—were greeted with exuberance. Cable went through its "blue sky" period, in which lofty plans painted an optimistic outlook for the information society. This era began in the mid-1960s but, "in one form or another, Blue Sky thinking would shape the business through its next thirty-plus years."[33] Considerable fanfare accompanied a demonstration of two-way functionality at the

1968 convention of the National Cable Television Association (NCTA).[34] By the 1970s, the Federal Communications Commission (FCC) proposed new rules permitting cable operators to import distant signals into the hundred largest US markets, contingent on some stipulations, including a requirement that new systems build two-way capacity into their plant.[35] These proposals spurred considerable prospecting. Perceiving limited demand for a service that merely conveyed broadcasters' outputs, analysts suggested that to "realize the full potential of cable," operators would need to offer interactive features that exploited the technology's varied capacities, among which electronic shopping was consistently listed.[36] At the 1971 NCTA convention, FCC chairman Dean Burch warned cable operators that they could expect regulatory disfavor if they did not augment their broadcast retransmission function by offering innovative services and experimenting with "such two-way operations as shopping from the home."[37] That same year, a report from the Rand Corporation admitted that remote shopping was "technically feasible" but not yet economically viable. "Remote shopping would be attractive to advertisers eager to stimulate impulse buying," the report acknowledged, but it was doubtful that subscribers would share advertisers' enthusiasm.[38] Despite such caveats, the promise of interacting with consumers via television was becoming a fixture in the definition and development of cable.

INDUSTRIAL PROSPECTING AND PREDICAMENTS

BUILDING A BUSINESS

Soon after two-way functionality was demonstrated at the 1968 NCTA convention, stakeholders began asserting visons for an interactive television business. Already by 1970, Teleprompter Inc., the largest multiple system operator (MSO) at the time, recognized itself as a "broadband communications company."[39] Similarly, the vice president of Cox Cable urged the industry to exploit television's "unused capacity" by deploying shopping services and other applications.[40] Teleprompter's president saw two-way services as a beachhead for cable operators to establish themselves in the data transport and market research businesses, and he acknowledged "a tremendous opportunity for merchandising of goods."[41]

This tremendous opportunity became part of the cable industry's value proposition. In negotiating a franchise agreement with New York City in 1970, Teleprompter touted its development of "armchair shopping."[42] The next year, Telecable Inc. tested a home shopping application that featured live presentations from a Sears, Roebuck store and used an advanced home terminal to let "the housewife . . . make choices on the spot by punching the appropriate buttons."[43] In 1973 Theta Cable and American Television & Communications tested interactive services, including shopping, in California and Florida, respectively.[44] Warner Cable launched its pathbreaking QUBE system in Columbus, Ohio, in 1977. Among other features, QUBE let viewers "order merchandise displayed on the screen, and even pay for it—by punching out credit card number and other required information."[45] In 1979 a former NCTA president started a cable company to realize "the medium's unfulfilled technological promises," using interactive services such as shopping to expedite cable's maturation as a general information infrastructure.[46] Cox Cable followed suit by the early 1980s, designing Index, a two-way data exchange system that facilitated banking and shopping.

Although these services were costly and slow to materialize, futuristic promises became strategic resources for cable operators and the municipal governments that authorized system building. Interactive applications were bargaining chips in pivotal franchise negotiations during cable's urban expansion in the 1970s and 1980s. According to L. J. Davis, "Nothing, absolutely nothing, won the hearts and minds of the targeted cities like interactive television."[47] QUBE helped Warner Cable secure franchises in several large markets, including Dallas, Houston, Milwaukee, Pittsburgh, and Cincinnati. Reporters identified Index as a decisive factor in Cox's winning bids in Omaha and New Orleans.[48] Mile Hi Cablevision obtained a franchise in Denver with a proposal that included "full interactive services, including home security, shopping, and banking."[49] And all six bidders for five franchise areas in Chicago promised home shopping and other interactive services.[50] Within just a few years, however, most of this prospecting proved more fanciful than feasible, with "beleaguered operators . . . trying to get out of agreements to provide extravagant services and facilities."[51] As cable companies postponed or abandoned the interactive offerings touted during franchise negotiations, critics alleged that the main purpose of interactivity was bargaining leverage.[52] Interactive ambitions thus helped draw the map of cable service in at least two ways: they influenced franchising decisions that granted lasting incumbencies, and they created conditions for opportunistic MSOs to expand by absorbing overextended cable systems that had promised more than they could afford.

Well-publicized misfires helped rein in this speculation, and as the industry found its footing with satellite interconnection and a national policy set by Congress in 1984, cable become a reliable investment without resorting to exotic fantasies. Some of the energy surrounding electronic direct marketing and home shopping gravitated toward commercial online services such as CompuServe and Prodigy (a joint venture of Sears and IBM), accessed via home computers, as well as short-lived videotex services, which used television sets but tended to be associated with newspaper companies. By the end of the 1980s, a movement by telephone companies to market television services renewed shoppable TV's salience in the narrative of media convergence and its currency as a strategic asset in system building.[53] Lobbyists for the telecommunications industry used the promise of interactive services to justify deregulation, and after legal victories and favorable rule making, telephone companies announced plans to build video distribution systems. Some critics considered this a ruse by regional Bell operating companies trying to enter the long-distance telephone market, but these designs were nevertheless elaborate, expensive, and consequential.[54]

GTE, the largest independent telephone operator (which became Verizon after its acquisition by Bell Atlantic), began testing a home shopping portal called Main Street in several markets in 1988; it then embarked on a plan to construct a fiber-optic–coaxial cable infrastructure for interactive services in California.[55] By 1994, GTE had deployed Main Street near Boston and anticipated acquiring up to seven million new customers for its interactive products over the next decade.[56] Elsewhere, Ameritech won franchises throughout the Midwest and earmarked $29 billion over fifteen years to build hybrid fiber-optic–coaxial cable systems capable of supporting interactive shopping.[57] US West developed an "interactive mall" showcasing merchandise from Virgin Records, Nordstrom, J. C. Penney, and Ford. The venture's executive vice president distinguished it from traditional home shopping: "We are creating short-term television with an impact, and it is important to remember we're in the business of direct marketing." Planning to "combine entertainment and electronic retailing," US West imagined a future in which digital marketplaces would learn, predict, and cater to users' viewing and shopping habits.[58] Meanwhile, *Variety* called Bell Atlantic "the most aggressive" entrant into television among the regional Bell operating companies.[59] It planned to spend $15 billion between 1993 and 2000 to equip 8.75 million homes with five "killer applications," including home shopping and direct-response advertising.[60] Larry Ellison, CEO of Oracle, the database software firm managing Bell Atlantic's system, expressed his vision in terms of frictionless impulse buying: "You're sitting there watching the ABC News, an ad from Time Life comes on and suddenly you've got an opportunity to order the entire works of Nat King Cole on CD. One click of the button and it's yours."[61]

Although most of these ventures ended in retreat from video provision, in the discourses about them we see vivid impressions of the modern internet, including high-capacity servers that store digital video for on-demand retrieval, easily navigable retail portals, click-to-buy shopping, dataveillance, and behavioral ad targeting. Telecoms' efforts to provide these information and entertainment services, evoking long-standing dreams of interactive television, influenced the building, financing, administration, and regulation of America's information infrastructure.

Cable operators knew a good act when they saw one. To outdo telephone companies, exert a competitive advantage over fledgling direct-broadcast satellite businesses, and search for new billable services to offset the subscription rate regulations Congress imposed in 1992, MSOs invested billions in plant upgrades and plotted their own adventurous schemes. In 1993 Time Warner announced plans for its Full Service Network, which is best remembered as a very expensive demonstration of video-on-demand. But another widely publicized feature let viewers use their televisions to order from Pizza Hut and make purchases from an "interactive digital shopping mall."[62] The company claimed to be transforming the television into a platform for multimedia and commercial applications (figure 7.1). At the same time, Cox Cable implemented a large-scale test of an interactive offering in Omaha, Nebraska, that included home shopping.[63]

Even though these experiments inflicted financial wounds and exposed the gulf between rhetorical hype and the viability of these plans, they flashed enough potential to excite an important audience of investors. Microsoft's $1 billion investment in Comcast in 1997 signaled that Bill Gates expected cable operators to build the interactive video systems he had imagined in his recent book. *Broadcasting & Cable* applauded Gates's financial blessing as visionary: "In the past 12 months, maybe in the history of cable, no other single event has done more to highlight the industry's potential and endorse its technology."[64] The billionaire who hoped television would become a shoppable, moving-picture catalog helped position

FIGURE 7.1
This promotional display for Time Warner's Full Service Network, exhibited in the equipment archive at the Cable Center's Barco Library, reiterates the vision of convergence via interactive TV that surfaced in 1970s discussions of a "wired world." (Photo by the author)

cable operators to become the dominant providers of internet service in the United States. And Microsoft saw a chance to control a critical bottleneck: the electronic equipment that would deliver a digital revolution into customers' homes.

BUILDING A BETTER BOX
Shoppability was a handy lodestone for excitement about the next generation of set-top boxes. STBs had emerged in the late 1960s as converters that enabled TV sets to receive cable transmissions from frequencies outside VHF and UHF bands. They became important instruments of control when subscription channels, tiered services, and pay-per-view offerings required cable operators to scramble signals and discriminate among customers using "addressable" systems controlled by computers at the operators' headend facilities. By the early 1970s, the cable industry recognized home

terminal equipment as a means of furnishing interactive services such as point-to-point merchandising, and venture capitalists began to exhibit interest in cable hardware.[65] Anticipation of these possibilities simmered for the next two decades before coming to a boil in the 1990s, when advances in digital technologies seemed to herald the arrival of a long-awaited future. Interactive direct marketing was presented as a central axis of revolutionary change.[66] A 1996 product brochure from leading STB maker General Instrument drew on the same blue-sky imagery that Time Warner used to promote its Full Service Network. General Instrument boasted that the "addressable intelligence" in its cutting-edge STB "turns television into a video store, shopping mall, library, brokerage house and more."[67] Despite nagging doubts about viewers' appetite for these new TV experiences, outfitting STBs with microprocessors and internet modems to support a suite of multimedia and marketing services was a highly publicized priority for leaders in the cable, software, and electronics industries.

The STB was being redesigned to materialize the hopes of an interactive future. It became a locus for the collision of television and networked personal computing. As an article in *Broadcasting & Cable* put it, "the set-top box is a nexus where different technologies can come together and generate new revenue."[68] The 1990s witnessed a gold rush into STB markets, as many people expected these devices to be the main consumer gateway to an "information superhighway."[69] The ranks of companies designing and manufacturing the apparatus for multichannel television—led by General Instrument and Scientific-Atlanta—swelled to include giants from the computer industry, notably Microsoft, Intel, IBM, Apple, and Hewlett-Packard.[70] These firms built digital components for cable, direct-broadcast satellite, and telecom operators that requested electronic shopping and other interactive capabilities.[71] Tech start-ups likewise rushed to design shoppable applications. For example, Wink Communications, an early standout in television commerce, enabled viewers to click their remotes to buy CDs from musical guests on *The Tonight Show with Jay Leno*.[72] According to the title of one consulting report, these "t-commerce" initiatives were "Turning TV Sets into Cash Registers."[73]

To some, this marked "a whole new phase of the cable and computer industries," positioning television companies to tap "the home retail market, which may be worth hundreds of billions of dollars every year."[74] TCI's executive vice president of ad sales suggested that advertising within the "direct-response environment" enabled by digital, addressable STBs constituted "a whole new way of using television."[75] By the late 1990s, the perceived affordances of digital cable systems had aroused considerable excitement: "The broadband pipe is primed and advertisers are pumped up about the prospects of translating the PC 'click through' to the TV."[76] Microsoft in particular looked to make itself an indispensable intermediary within this stack of technologies by using its operating system to control the STB.[77] Beyond licensing its software, Microsoft hoped to capture fees on e-commerce transactions, which were expected to skyrocket. Even in 2002, when internet browsing had settled around desktops and laptops, *Fortune* called the STB "the most valuable square of real estate in America."[78]

Analysts responded to the convergence of television and personal computing with forecasts of rapid market growth. One study projected that interactive television shopping revenue would total $4.3 billion in 2005, with "the bulk of this buying" executed directly by remote control.[79] Other estimates were even grander.[80] But this optimism was disappointed. First of all, building an STB to actualize the dreams publicized by cable operators was difficult and expensive.[81] The set-top terminals running Time Warner's Full Service Network reportedly cost $5,000 to $10,000 apiece.[82] Even in the more modest systems planned in the mid-1990s, the price tag for next-generation STBs ($400–$1,000) was well above the perceived threshold of viability ($200–$300).[83] Activating the capacity implied by digital STBs also required outlays for home installations and upgrades to headend facilities, and cable operators were reluctant to replace equipment that was still being amortized over its expected life span. Even for MSOs with enough scale and cash flow to absorb these expenditures, the process was complicated by the patchwork nature of their footprints, which they had built by acquiring local cable systems whose facilities varied in age, quality, and compatibility. These pressures depressed STB orders, kept manufacturing costs high, and discouraged enterprises that required mass deployment of digital equipment.

Furthermore, as an infrastructural technology, STBs were entangled with actors, institutions, and interests across industries and sectors. While shoppable television excited imaginations, STBs were designed to receive and control access to video content. Building an interactive storefront occupied an outsized place in discussions about the future of entertainment, in comparison to more pressing concerns such as program licensing, signal security, and digital video standards. Futuristic services made for good publicity, but filling the bandwidth unleashed by digital compression with more programming was a safer bet for cable operators. Moreover, the coordination needed to stabilize a network of interoperable devices and protocols with enough scale to satisfy national marketers was undermined by proprietary dispositions among system operators, equipment makers, and software developers, each vying for the monopoly windfalls to be had if their products or services became infrastructural standards.[84] As one article admitted, TCI, Time Warner Cable, and Microsoft "have spent billions in pursuit of this Holy Grail, with next to nothing to show for it."[85]

Consumers, meanwhile, encountered clunky interfaces, a dearth of attractive interactive content, and invitations to engage with TV in unfamiliar and even unappealing ways. Asking viewers to interrupt their entertainment to go shopping was bound to annoy people. For their part, advertisers and marketers continued to face a promising but largely incoherent technical and administrative environment. Although the technologies existed to facilitate shoppable television, that potential had not been realized in industrial process or cultural habit. A cable executive with an apparent knack for understatement said, "Sometimes the dreamers dream faster than the implementers."[86]

DREAM FAST AND MAKE THINGS

Despite these stumbles, the dreamers trudged on. In 2000 Josh Bernoff speculated that viewers would soon be able to purchase Jennifer Aniston's sweater while watching *Friends*. This had an intuitive appeal for marketers. *Friends* was a cultural phenomenon, and Jennifer Aniston was at the pinnacle of fame. But the details were sobering. For example, nobody could work out how to share sales revenue among the garment maker, the t-commerce service provider, the broadcast network, the affiliate station, the cable or direct-broadcast satellite operator, the show's producers, and the actor herself. As skeptical observers perceived, with "too many fingers in the t-commerce pie . . . the economics of Jennifer's sweater quickly unravel."[87] One critic called the idea "pure rubbish," adding, "I don't think anyone will want to watch *Friends* to buy a sweater."[88] Bernoff soon admitted that "t-commerce expectations have been way overblown."[89] Nevertheless, Jennifer Aniston's sweater became a touchstone for television's direct-marketing future. A 2008 article in the *New York Times* called "Rachel's sweater" a "catchphrase . . . for what devotees of interactive television are trying to accomplish."[90] It was "blue skies" for a digital age.

The potential to enable impulse buying—minimizing the steps between introducing and consummating a purchase opportunity—was already apparent by the 1970s and was a motivating factor in designing cable systems that could support impulse pay-per-view. The vision articulated by Gates and Bernoff further captivated imaginations, aided by television's history as a showcase for attractive goods, services, and lifestyles. Shoppability rolled right off the tongue for commercial media. "Talk about impulse buying," exclaimed *USA Today* in 2005. "You're watching your favorite cable channel and admire a product on the show. With a few clicks on your TV remote, it's yours."[91] Enthusiasm swelled as digital STBs were installed in thirty-eight million homes by 2008, and interactive television, according to a Cablevision executive, was no longer the "wave of the future" but the "wave of now."[92] The *New York Times* observed that "cable companies are starting to slowly move" toward "the promise that consumers could instantly buy Jennifer Aniston's sweater on 'Friends.'"[93]

Versions of this trope were inherited to characterize initiatives that let television viewers use "second screens" (laptops, tablets, and mobile phones) to buy products related to programs and advertisements.[94] As more TV viewing occurred within reach of internet-enabled devices, advertisers, programmers, and video distributors leveraged this connectivity to "turn any network, app, multichannel provider and TV set into a kind of home shopping network."[95] As early as 2006, a company called Delivery Agent began operating an online tour that turned the set of the TV show *Desperate Housewives* into a shoppable showroom.[96] Shazam, a mobile app designed for music discovery, was adapted to "reinvent the 30-second spot" by allowing "viewers to buy products from mobile devices."[97] In 2016 A+E Networks produced "the first 'fully-shoppable' TV series" in which every item featured on a home improvement show could be purchased from Wayfair.com.[98] Just a few years later, AT&T tried to cash in on its reach as a large mobile carrier by targeting subscribers' phones with advertisements, coupons, or shopping opportunities related to what they were watching on TV. Brian Lesser, then CEO of AT&T's advertising business, explained, "Imagine, you're watching content and instead of us interrupting the content with a traditional commercial break, we can show an icon on the screen that indicates to you that there might be a mixed reality experience where you can get more information about the car you just saw or the dress you just saw."[99] Before joining AT&T, Lesser tried to implement a similar program at Xaxis, WPP's programmatic platform.[100]

While these second-screen initiatives introduced another device beyond the TV and its remote control, they contributed to the project of making video content shoppable. Increased connectivity to digital marketplaces helped sustain the hope that selling merchandise directly through television devices was finally becoming mainstream, especially as firms with core competencies in electronic retailing, such as Amazon and Walmart, assumed larger roles in video distribution.[101] Having overcome the barrier of "getting that living room connected," Mike Fitzsimmons of Delivery Agent told CNBC in December 2015, "this idea of having a tethered experience between the content that you're viewing and the ability to purchase is becoming a reality."[102] A year earlier, when Fitzsimmons announced that his company would deliver a shoppable H&M advertisement during the 2014 Super Bowl, he said t-commerce had realized at last "the potential associated with buying Jennifer Aniston's sweater."[103]

Less than a year after Delivery Agent filed for chapter 11 bankruptcy protection in 2016, Fitzsimmons started another company, Connekt, to keep chasing the dream. According to a fawning company profile in *Broadcasting & Cable*, "the technology appears to have finally caught up to the idea." By bringing t-commerce to internet-connected devices, including smart TVs and Roku boxes, "Connekt appears to have figured out how to turn viewers into instant consumers with the touch of a remote button."[104]

LET BUY BUTTONS BLOOM

As one writer put it recently, "frictionless consumption is big business's wet dream."[105] And it has been for a while. Shoppable ambitions course through the history of commercial media. From radio to cable to early online services like Prodigy, marketers and retailers have tried to turn new means of communication into more intensive and personalized tools of commerce. "Buy buttons" have bloomed all across the internet. A handful of intermediaries bring brands and content creators together and manage the details of compensation and order fulfillment that plagued shoppable TV. Influencers on Instagram and TikTok help merchandise the clothes, accessories, elixirs, cosmetics, and experiences they demonstrate in sponsored posts, which are often designed to give the impression that audiences can own a piece of the influencer's lifestyle or persona. Online platforms, including Amazon, are scrambling to recruit and groom popular influencers and surround them with the tools needed to sell the set dressings of their lives to loyal followers. The whole field of influencer marketing is too varied to summarize here, but much of it advances the familiar project of using celebrities and parasocial relationships to sell. Influencers have gravitated toward practices of authenticity, managed intimacy, and microcelebrity, yet the body types, beauty standards, and visual styles that marketers tend to invest in, as well as the top-heavy concentration of wealth and fame, exhibit as much continuity as change.[106]

It is a fact of life that more and more of what we see online is available for sale. Huge swaths of the internet have become the interactive catalogs imagined by futurists past. As the COVID-19 pandemic forced people to do more shopping and socializing online, marketers and platform companies seized on the crisis to further entrench the shoppability of social life and enclose social commerce within their garden walls. It is a power play that leverages an active history. Journalist Shoshana Wodinsky brings that lineage into focus with the headline "TikTok Wants to Be QVC for Teens."[107]

CONCLUSION

This history of shoppability shows how commercial expectations framed problems and potentials for system builders, including the promoters trying to make new technologies meaningful to various publics. From the late 1960s through the 1990s, as the cable television and telecommunications industry worked to define its place in a convergent media landscape, interactive home shopping starred in seductive stories about the future of entertainment and information services. As an alluring and long-imagined consumer service, interactive shopping was a useful publicity device for a political-economic process of media convergence and marketization; it shimmered on the surface of a wave that swept through business and policymaking. These stories influenced the contours of cable's footprint and regulatory framework and were built into hardware and software as part of an effort to position digital STBs as the domestic portal to a version of the information revolution inflected by mind-sets and structures from advertising and direct marketing. Shoppability stirred marketers' fundamental interest in using technology to accelerate the circulation of commodities. Transactive television catered to the logic of commercial media, stated succinctly in *Advertising Age*: "The business of marketing and the business of entertainment are fundamentally about the same thing: Turning audience attention into commerce."[108] The promise of selling Jennifer Aniston's sweater never fully unravels because it weaves the yarns spun from restless dreams of making any medium into a marketplace.

Specific visions of shoppable television developed more slowly than many hoped, for a variety of technical, economic, cultural, and political reasons. But it is hard to deny that the efforts to construct shoppability as an affordance of interactive media have profoundly influenced the meanings and the commercial and cultural practices associated with today's digital environments. Within the last decade, and accelerating even more in recent years, companies have spread shoppability across all sorts of media content, devices, platforms, and applications. From digital magazines to ads and entertainment on YouTube, Instagram, and TikTok, the notion that media are marketplaces is beyond dispute. Amazon now offers a reverse-image search feature, StyleSnap, that makes almost any social media post shoppable; consumers can upload a screenshot to find the same or similar products on Amazon, and affiliates of the Amazon influencer program get credit for generating sales.[109] With the near ubiquity of mobile phone cameras and image-recognition software powered by artificial intelligence, some marketers even turn physical surroundings into shoppable spaces. Visual search tools such as Google Lens and Lykdat let users capture images of items worn by passersby and then find online merchants selling whatever caught their eye.

The linking of media usage to shopping arouses a desire to *know* that viewers bought what they saw advertised. The next chapter examines accountability, an affordance that the advertising industry looks for in almost every new information technology that enters its field of vision and influence. Because accountability spans the whole history of advertising's calculative evolution, the chapter begins back at the midcentury inflection point, when management technique promised marketers a world they could count on.

8 VIVE LE ROI! ACCOUNTABILITY AND THE PULLING POWER OF ATTRIBUTION

> The problem of measuring advertising's ultimate effectiveness is as old as advertising itself. One hopeful step toward a solution has recently been taken by companies employing certain mathematical and analytic techniques loosely defined as operations research.
>
> —*Minutes of the First Meeting of the Operations Research Discussion Group*

Looking at the impact of management science on marketing up to the mid-1950s, a professor at the Carnegie Institute of Technology lamented a considerable lag in comparison to its impact on production. Melvin Anshen said the factory setting lent itself to observation and control, and applying "the science of cost accounting" in manufacturing generated valuable information that "was a further incentive for rationalizing management." The situation in marketing was messier: relationships between variables were harder to isolate from confounding factors, and some of the most important variables and interactions were nearly impossible to measure or manipulate. Except for direct marketers and some grocery and drug makers that subscribed to sales tracking services, "manufacturers do not know how to establish direct cause-effect relationships between sales promotion outlays and sales to ultimate consumers." They were left to rely on "measures of indirect relationships," such as evidence of advertising exposure, or to draw conclusions from "the raw information on what was spent and what was sold, with such a gossamer bridge between the two as the interested parties may be willing to construct." That hardly seemed like a sturdy bridge to the future. The mounting weight of marketing costs compounded pressures to exploit the "outstanding potential contributions" of optimization techniques such as linear programming. Crucially, though, Anshen observed a major impediment: "One dominant influence in the marketing process—the consumer—is outside of management's direct control and is only partially, and until now usually unpredictably, susceptible to manipulation and influence."[1]

Operations research excelled at orchestrating factors of production and distribution that responded faithfully to directives and immutable laws. But people existing outside an organization's scope of command did not always conform to management's plans. The whole project of using operations research and management science (OR/MS) to optimize advertising implied ambitious designs to account for consumers' economic value and their expected and confirmed behaviors. Management science machines demanded and enacted calculable consumers whose attributes, activities, relations, and probable futures could be materialized as data and represented numerically in models. Rendered in these ways, by drawing on practices such as cost accounting,[2] complex marketplace actors would apparently become more legible and manageable. According to Sigfried Giedion, efforts within scientific management to precisely measure industrial motion were aimed at "giving a full account" of workers' movements.[3] OR/MS promised progress toward a related aspiration: to precisely measure consumers' movements and give a full account of advertising productivity.

Advertising's calculative evolution is defined in part by the expansion of models and measurement capabilities to accommodate more consumer behaviors. Marketers have essentially tried to extend managerial observation and control over the sphere of consumption. The goal is to account for the outcomes of advertising investments and the value of advertising opportunities, where accounting means not only recording and enumerating but also admitting those quantities into the frame of rational planning and action. "By enhancing the inventory of relations and events to be taken into account," Michel Callon explains, "marketing tools promote calculations which constantly involve more and more elements and relations."[4] Advertisers have dreamed of not simply paying to put messages in front of potential consumers but rather reorienting buying procedures around the verifiable results of their investments. To put it more formally, advertisers want the objective functions in their optimization models to directly represent consumer purchases or company profits, rather than proxy measures such as audience ratings or shoppers' attitudes toward a brand. "If media planning is ever to be put on a more scientific footing," a 1964 report by the Advertising Research Foundation (ARF) urged, "additional ways must be found to fasten the connection between media and ultimate sales."[5] Advertisers enlisted operations researchers for exactly this mission—to reinforce the "gossamer bridge" between advertising and its measurable results.

TAKING AND GIVING ACCOUNTS

OR/MS established its place in the marketing field by turning the determination of advertising effectiveness into a matter of disciplinary concern and consulting expertise. The process of making claims about "the 'chain of causation' from selling stimulus to buying response," as one statistician put it in 1957, is what the industry now calls "attribution."[6] It is an approach to calculating return on investment (ROI) whereby specific marketing actions are credited for their discernible contributions to marketing outcomes. The will to attribute sales to particular advertising decisions and events has channeled energies toward the present state of the art, where advertisers and adtech intermediaries build campaigns and transactions around specified behavioral outcomes. Attribution is the crowning achievement of OR's original promise to optimize ROI. And it has been a siren song for consumer surveillance and datafication initiatives.

This chapter examines the persistent desire to measure consumer behaviors in response to advertising and to incorporate those measures into advertising's exchange relations and evaluative metrics. It is about *accountability* in a literal sense—the ability to take things into account. Advertising accountability has two interrelated vectors. One is to account for advertising effects in terms of ROI. The other is to account more precisely for the expected value of consumer populations, audience targets, or even an individual's momentary propensities. Both vectors depend on capacities to generate data, to process and analyze those data, to assemble new claims (about advertising effects, or how to classify and value a profiled consumer), and to legitimize those claims within interorganizational action. Accountability, therefore, is also rhetorical.[7] By documenting consumer behaviors and advertising effects more carefully, marketing professionals can ostensibly give more convincing accounts—producing evidence of optimal decisions made rationally.

One of the most forceful arguments promoting digital advertising is that it takes advantage of the record-keeping and identification capacities of internet servers, networked databases, analytics software, and tracking tools like cookies and pixels. Advertisers invest in digital media because they are more accountable—they datafy more behaviors and relations. This has been conventional wisdom for two decades. But the arguments and operations we associate with digital adtech capitalize on the particular historical development of accountability as an affordance. The envisioned potential to better account for the value of consumers and advertising investments has been constructed over half a century and attached repeatedly to new information technologies. Computer systems and mathematical models provided a powerful set of scientific, technological, and discursive resources for defining and venerating accountability in marketing. More than just recording reality, these means of accounting enacted realities in particular, strategic ways.

Apostles of management technique confronted advertisers' attribution anxiety—their long-standing fear that some unknown portion of ad spending is wasted—by suggesting that technoscientific breakthroughs were making it possible and, indeed, imperative to account for ROI. Computerization and OR/MS necessitated more consumer data and increasingly frequent and intensive forms of market analysis. Marketing's emergent class of mathematical elites tried to harness new technologies to their vision of a world where more spaces and devices would be instrumented to collect behavioral measurements and where those measurement instruments would be integrated into adaptive management and decision support systems. Perhaps as much as the desire for better targeting, the dream of determining advertising effectiveness has motivated marketing professionals to position themselves where they could witness more scenes from consumers' lives.

An expansive vantage point is essential for attribution. Isolating the unique effects of specific advertising events requires both close observation of the causal chain between exposure and eventual marketplace action and a careful accounting of the many other factors that can influence the relationship between advertising and sales. In other words, a credible attribution claim depends on evidence that an ad was served, seen, and acted on, as well as evidence that helps rule out alternative explanations. Not surprisingly, then, attribution is a pesky knot to untangle.[8] Researchers funded by the Leo Burnett agency began an article in *Applied Economics* by admitting, "Analysis of the influence of advertising on sales is a difficult and dangerous undertaking."[9] Some researchers concluded that it was hopeless. "Over the past generation," Leo Bogart huffed in 1966, "the research world has been strewn with the bleached skeletons of innumerable unsuccessful attempts to measure accurately the specific effects of a specific advertising campaign."[10] Importantly, though, the dream never succumbed to frustration. Quite the opposite. Failure to find conclusive evidence of advertising effects only motivated efforts to look harder—and to hire more skilled technicians. Frank M. Bass, one of the researchers funded by Leo Burnett, reaffirmed his commitment a decade later when Stanford University hosted the first OR/MS conference on marketing measurement and analysis: "Despite the difficulty and dangers of formal analysis of the effects of advertising and promotion upon sales, the profit potential from the application of modern statistical methods and management science systems is very great."[11]

For the advertising industry, accountability is perhaps the most seductive affordance that can be attached to new technologies. Its appeal is woven into the logic of selling the American people. Faith in the power of advertising has been a fixture of US corporate culture. As a game theorist working for M&M Candies put it in 1959, "every executive knows that advertising has an effect on the sale of his products despite the fact that he may not be able to measure this effect quantitatively."[12] This performative belief in advertising effects is existential. The whole massive advertising industry presupposes that advertising works, and thus a large professional community is committed to the idea that advertising produces sales and influences behavior. Vast assemblages have been built around the desire to account for these relationships, which are fiercely assumed to exist but nearly impossible to confirm.

The difficulty of determining advertising effects has helped justify strenuous efforts to monitor and identify consumers, to amass analytical power for parsing interrelated variables, and to leverage precise forms of discrimination and targeting. As observers have acknowledged for decades, uncertainty about whether advertising works has left marketers vulnerable to commercial research and data providers that promise to bridge this knowledge gap, often with answers that are pleasing but hard to adjudicate. To put it bluntly, the causal effects of advertising may be practically unknowable, but since that unknowability cannot be openly tolerated, the challenge of sifting through the mess of reality to mine evidence of attribution is an important source of motivation and power. It gives a strong and persistent warrant for investing in datafication, analysis, and technical expertise, and it supports today's gargantuan markets in surveillance, data, and analytics. One of the strongest competitive claims made by walled gardens such as Google, Meta, and Amazon is that they know, better than anyone else, what advertisers are getting for their money and that they can tune their management science machines to optimize ROI. The commitment to determining advertising effects has itself had profound effects.

CASH REGISTERS AND BALANCE SHEETS

It would be hard to overstate the fixation on ROI in certain parts of the advertising business. At a 1965 seminar for media buyers and sellers, the director of marketing at the Doyle Dane Bernbach agency emphasized this point by redefining the nature of the "client" being served in advertising. "It's neither a product nor an advertising manager nor a company president. *The client is invested capital.* The capital is invested to make a profit and that is exactly the name of our game. It is the function of our advertising to produce profit for the corporation." Calculating the effect that a change in ad spending would have on profit, he said, was among "the greatest open questions in our business today."[13] A business professor later wrote in *Management Science* that the relationship between advertising and sales was probably "the most researched area in marketing."[14]

The roots of this outlook run throughout the modern history of US advertising. Ever since advertising became understood as a mediated form of salesmanship, Daniel Pope explains, "advertising people of all stylistic bents have agreed upon the purpose of the work: the task of advertising is to sell. The only legitimate measure of success is at the cash register." This "credo" not only unites management and creative production around a common objective; "it is a fundamental element of continuity in advertising from the beginning of the [twentieth] century until today."[15] This is the culminating logic on one side of selling the American people—where advertising tries to accelerate the circulation of commodities. But on the other side—where media companies package and sell evidence of audience attention—the business has been organized around practical compromises.

Attention, audiences, and ratings are all proximate constructions of the influence advertisers are trying to buy. Dissatisfaction with exposure-based audience measurement endured throughout the broadcast era. Agitation for more behavioral or cognitive means of evaluating advertising opportunities came both from advertisers and sometimes from broadcasters that wanted to tell a better story about what they were selling than ratings permitted. Responding to publicity about a new approach for measuring radio and TV audiences in the late 1940s, the executive vice president of a Massachusetts station wrote, "I don't think that any radio research technique can be termed revolutionary until a method is developed which will determine and disclose the impact of radio on the mind of the listener. . . . The success of a program can be known only by its effect upon its hearers, whether its purpose is to entertain, to sell merchandise or both."[16] Thirty-five years later the former director of marketing research for Lever Brothers reiterated that ratings data were merely *the surrogate for what advertisers want to have but do not yet have.*" After all, he said, it is "the function of media planning to optimize the effectiveness of the media, not only in terms of reach and frequency but more importantly in terms of response."[17]

Advertisers wanted audience measurement to get close enough to consumers to hear the cash register ring. But it was easier and cheaper to produce evidence of advertising exposure than evidence of its effects. And exposure, though still contested, was a simpler construct to standardize for pricing and exchanging commodity audiences. To the extent that advertising effectiveness could be brought into the picture, most of the early social-science research tried to measure the impact of advertising messages on consumer attitudes or memory, rather than to attribute purchasing behaviors to those ads. A large field of mostly psychological research looked for evidence that audiences recalled and retained commercial communications. These were some of the questions that gave rise to an industry in opinion polling and the social construction of a mass public.[18]

Operations researchers, and the agency departments they worked with, demanded information about audiences that was both detailed and standardized—data suitable for modeling and computation. Bogart wrote that OR's "huge appetite for data" helped create "a whole new priestly caste: the syndicated media researchers."[19] But operations researchers were seldom satisfied with the exposure-based ratings sold by Nielsen and others. Furthermore, they were not especially interested in whether and how people absorbed advertising messages. "Nobody will argue seriously that the ultimate goal of advertising is to communicate rather than to increase sales or the profits of the company," explained Michael Halbert, an OR specialist in DuPont's Advertising Research Department. "You might move to a communication measure because you would like to measure sales but you can't."[20] Companies tried to optimize communication objectives, Halbert said, because management could easily control and observe the relationship between an independent variable like ad spending and an outcome variable like exposure. By contrast, the sales response to advertising is a function of many interacting and possibly unobservable factors. "If you use sales as the measure of effectiveness, you have additional variables you don't have if you use impressions or attitude change or favorableness toward the product, etc. . . . It's easier to build a model if your measure of effectiveness is change in attitude or information: on the other hand, as is always true, it's less useful."[21] MIT professor Ronald A. Howard stated the matter frankly at a 1962 meeting: "*intention* to buy doesn't go on the balance sheet, only the sales."[22]

Operations researchers were invited into advertising with a mandate to apply their unique skill set to exactly this problem. A professor at Princeton University observed that businesses were increasingly turning to "economists and statisticians for analytic advice concerning the effectiveness of their advertising effort." This pattern seemed to him "a natural accompaniment of the noted successes achieved by operations researchers in other business problems."[23] Already by the 1940s, one mathematical astronomer was using what he would later describe as OR to measure the "pulling power" of department store ads, or "*the amount of sales produced per dollar of expenditure.*"[24] These techniques flowed into a current of work aiming to refashion marketing as a more quantitative science. "The principal hope for developing more rational promotional plans," wrote Joel Dean of Columbia University, "lies in improved methods of measuring the effects of advertising." He acknowledged, however, that "relating outlays to the sales they produce is a formidable problem and calls for elaborate statistical attacks."[25] Pollster Sidney Hollander Jr. thought these statistical attacks would allow marketers to treat advertising as a measurable investment, rather than setting budgets arbitrarily or as a percentage of past sales. He hoped tools such as "graphic multiple correlation" and "factor analysis" would enable "the rationalization of the advertising appropriation to achieve the greatest net return for each dollar of expenditure."[26] He admitted, however, that the data needed to perform these calculations were lacking, so he advocated for the refinement and extension of market research.

There was widespread agreement on that point. Outlining the future for "behavioral management science," Samuel P. Hayes Jr. emphasized in 1955 that a "science of consumer relations management" would require intensive observation and analysis of consumer activity. Hayes, who had previously worked as a researcher at Young & Rubicam and Dun & Bradstreet, wrote, "Much more will be known tomorrow than is known today about consumer motivation, buying habits, and susceptibility to particular advertising media and themes; but, they will by no means obviate the necessity . . . of continuous collection and evaluation of data."[27]

Advancing a science of consumer relations management was thus seen as a problem of observation and classification. If attribution was to be more than a gossamer bridge between gross expenditures and total sales, then buying behavior needed to be documented and assessed more carefully— marketers needed to take more exact measurements of consumers and their value to the firm. A canonical essay on market segmentation noted that the strategy depended on "the maintenance of a flow of market information" and "the full utilization of available techniques of cost accounting and cost analysis."[28] Management scientists encouraged advertisers to use cost accounting to identify which consumers, markets, and sales were actually profitable and which were a drain on marketing expenditure.[29] As early as the 1930s, marketing theorists had "advocated that promotion costs be allocated to individual customers on the basis of the amount of gross profits obtained from them."[30] Charles H. Sevin of the US Department of Commerce suggested that better accounting of marketing costs would fuel "a definite acceleration of the tendency toward selective distribution. Instead of merely striving for an ever increasing sales volume regardless of its profitability, the aim of marketing will shift toward 'selling the right products, in the right quantities to the right people in the right locations.'"[31] Writing with William Baumol in 1958, Sevin suggested pairing "distribution cost analysis" with mathematical programming. By calculating the "best" option for allocating marketing investment using "the standard optimization technique, i.e., differential calculus," the "businessman" could make marketing decisions that were "virtually guaranteed to increase his profits."[32]

"FORMULA'D" FIGURING

By this time, advertisers and their agencies were working with operations researchers to put these techniques into action—examining advertising effectiveness and trying to formulate mathematical decision rules for setting advertising budgets and allocating expenditures.[33] They experimented with ad spending in different regions or at different times and then tried to discern the sales effects attributable to advertising.[34] General Electric began applying OR to the problem of measuring ROI in 1957. As a researcher there explained, "Applications of computers to the problem brought new hope. These electronic marvels make it possible to analyze masses of data in multiple combinations to see what relationships exist between the dollars spent on advertising and the dollars returned in sales. Businesses everywhere are 'playing with the figures' hoping a formula'd answer will be forthcoming. 'Spend "X" dollars on advertising and expect "Y" dollars in sales.' That's the hoped for equation."[35] Y&R described its work in this area as an effort to develop a "measure of the productivity of advertising."[36] Y&R's Kenneth Longman explained that "so many factors other than advertising play such a considerable part in producing sales that attempts to attribute sales changes to advertising can be hopelessly misleading."[37] With help from Wharton professor Wroe Alderson, Y&R researchers tried to break the impasse with brute-force statistics: "an elaborate sifting of all the possible influences upon sales." They produced answers and recommendations for the client, but the effort was not a resounding success.[38]

DuPont took this concept further than perhaps any other advertiser at the time. The petrochemical giant stands out as an early leader in using OR to study advertising and in orienting corporate planning around ROI.[39] DuPont's Advertising Research Department housed a staff of operations researchers that included Michael Halbert, the founding chair of the ARF's OR discussion group, and Malcolm McNiven, who later held senior management positions at Coca-Cola and Bank of America. These researchers tried to determine advertising effectiveness with iterative model building and experimentation: "this method involves first constructing mathematical models of how sales are achieved in the market, then making ad budget adjustments in test markets, using the results of the test to refine the model, testing again, and so on. With each step, the model should more accurately explain the effect of advertising and other factors on sales."[40] This model, once refined, would be used to "predict the future effects of changes in advertising." But despite the pricey computer equipment, DuPont's predictive power remained weak, a shortcoming attributed to "the practical difficulties in the way of feeding the oracle enough data."[41] Still, the possibility of discerning how changes in advertising policy impacted profits held enormous appeal.

Even a minor improvement in efficiency could mean substantial profits for a company spending $30 million on advertising to consumers and industrial customers in 1958.[42] DuPont later hired Simulmatics to continue this work,[43] and by 1963 the firm was budgeting $700,000 to study the effectiveness of its advertising.[44] DuPont's manager of advertising research told a meeting of the Association of National Advertisers in 1966 that the company believed these inquiries "should be capable of determining the contribution advertising makes to profits."[45] He also mentioned the company's progress on building predictive models.[46] An article published the following year reported that DuPont was able "to predict the sales response to a new advertising campaign within a few percentage points of the actual results."[47]

DuPont was one of a handful of corporations investing in the production of evidence—and evidential technique—of advertising effectiveness. These initiatives sent ripples beyond any single company. In addition to fueling evangelism for a more quantitative and scientific attitude in management cultures, OR/MS projects helped furnish new marketing databases. Advertisers' demands for rational media services spurred agencies to build "tremendously valuable" libraries of information.[48] In a 1956 memo about the recent restructuring of J. Walter Thompson's media operations, the agency described the mass of information accumulated in the service of some exacting clients: "The voluminous number of documents and charts that have been produced is a little staggering. . . . While we seem to have statistics running out of our ears for Ford, Scott [Paper], etc., this information has already had tremendous value in approaching media planning for all Thompson clients."[49]

An enthusiasm for mining insights from these information assets anticipated the now characteristic imperative to accumulate and hoard data in case something valuable might be discovered later. Wroe Alderson, director of the Management Science Center at the University of Pennsylvania, noted this impulse in 1963: "the constant pressure for better forecasts may occasion a deep regret for past [business] records thrown away. It is impossible to foresee all the possible uses of information which appears relatively useless today."[50] MIT's John D. C. Little also encouraged extensive data collection: "One of the great virtues, in my opinion, of getting into a program of continuous experimentation is that it imbeds in your past history excellent possibilities for analysis."[51] Furthermore, his enthusiasm for staging experiments to strategically datafy reality presaged the engineering disposition built into algorithmic adtech platforms: "Why not arrange things, if you are the president of the company, so that the world is easy to analyze instead of sitting back passively and letting the world go its own complicated way? Why not take your power of intervention and use it to maximize your own information state?"[52] This is a reasonable description of how Google and Amazon assert platform power in advertising today.

By the 1960s, "electronic computers had made vast stores of information available" to advertising managers.[53] These growing databases, and the statistical methods of extracting practical knowledge from them, enabled novel ways of sorting consumers according to their potential value. When Stanford University professor William Massy analyzed a data set of household purchases provided by JWT and the Market Research Corporation of America, he found that socioeconomic and personality variables were not helpful in predicting certain purchasing behaviors, such as brand and store loyalty. However, he saw promise in segmenting consumers into groups based on "promotional elasticities," meaning how sensitive their buying behavior was to advertising or price discounts. "I am convinced," he said, "that it's the dynamic response that is important for market segmentation."[54] Massy and other operations researchers wanted to use cost accounting and related methods to discriminate among different consumer populations—to recognize their "long-term value" and to distinguish them based on past purchases and "susceptibility to persuasion."[55] In other words, they were interested in classifying people based on their observed or inferred responses to manipulatable stimuli, rather than demographics. Their excitement was stirred more by emergent possibilities than the actual state of the art and the data, but wishful thinking out loud was itself a form of advocacy for new modes of accountability in marketing and media.

A WAVE OF ACCOUNTABILITY

Reflecting on advertising trends over the previous two decades, an editor at *Management Science* wrote in 1969, "There seems to be great advantage for media which can provide data on the audiences they reach." He speculated that this was related to the introduction of mathematical models for media planning. "Certainly models will tend to reward media with data," he explained, praising the fast-growing business of direct-mail advertising for its "extremely detailed" audience data.[56] This was a familiar observation. Direct marketing was generally perceived as more amenable to management science than was brand-image advertising. A business professor pointed out in 1957 that the use of research to gauge advertising effectiveness was much more successful when consumer responses were rapid and "readily traceable": "*The easier it is to trace response to marketing actions, the easier the problem for research.*"[57] Direct marketing gave managers the feedback they needed. As economist Julian Simon put it in 1970, "Mail order advertisers have an almost perfect measure of the effect of their advertising."[58]

Agencies promoted direct mail by highlighting its accountability. JWT positioned itself in the late 1950s as having "pioneered the development of two particular areas of the overall advertising activity—specialized selling, and the creation of measurable consumer action." In a pitch to Nationwide Insurance, a direct-mail specialist at JWT said mailings that solicited immediate returns "can be likened to a brokerage operation: for the investment of X dollars, the advertiser creates Y dollars worth of business, at a profit of Z dollars within a stipulated period of time."[59] He was promising a "formula'd" equation for advertising investment. Very soon, a high-profile movement gave these metrics and mind-sets a home at the top of agency management. For its champions, accountability was part and parcel of information technology.

In 1960 Marion Harper Jr., the head of McCann-Erickson, formed the Interpublic Group of Companies (IPG), integrating multiple marketing services within a conglomerate holding company. Harper proselytized accountability as a central value for advertising and marketing. "The whole idea of accountability is the wave of the future—even a wave of the present," he said in 1961. "It will now be possible, with the help of social sciences and mathematics, to measure advertising as a single influence, isolated from all the many variables involved in carrying on a business."[60] Harper even suggested linking agency compensation to sales outcomes, a possibility enabled by the growing availability of information.[61] He assigned researchers and information systems managers an important role in his vision of marketing administration. IPG convened an Applied Science Division, which, according to its manager from the mid-1960s, "brought together media researchers and operations researchers to create systems to improve the effectiveness of media decisions for the clients of all Interpublic agencies."[62] According to a 1967 profile of Harper, "His success with clients is grounded in the belief that advertisers seek above all else an assurance that their advertising dollars are being spent wisely. Therefore, the Interpublic organization places a great deal of emphasis on scientific approaches to advertising problem solving."[63]

Harper was ahead of his time in many respects. Agency holding companies later became the dominant pattern of industrial organization in advertising.[64] Throughout the 1980s and 1990s, the next generation of empire builders, such as WPP's Martin Sorrell, exploited the growth of "below-the-line" agencies that specialized in direct marketing. These firms were proficient in economizing and accounting for results. Importantly, "the below-the-line agencies were inevitably wedded to consumer and customer data, and technological and social developments were going to assure that data became more plentiful and powerful."[65] Harper agitated the industry to capitalize on information technoscience, and the diffusion of his conglomerate strategy intensified advertising's culture of accounting when mergers and acquisitions thrust agencies into the maelstrom of financialization, debt, and commitments to shareholder value.

Harper's plea to make agency compensation dependent on advertising effects also arrived too soon, running up against the existing commission system. Agencies earned almost all their revenue from what amounted to a refund by media companies—15 percent of the cost of the media inventory an advertiser bought stayed with its agency. This practice eventually fell out of favor, partly due to the interest in measuring advertising effectiveness. The commission system put agencies in an uncomfortable position with respect to advertisers' designs for setting optimal budgets. OR/MS investigations could, and sometimes did, conclude that advertisers were spending more money than they should.[66] Not surprisingly, agencies resisted recommendations to reduce ad spending, since that cut directly into their commissions. To resolve the conflict, some advertisers made new arrangements to incentivize efficient buying, rewarding their agency if spending decreased but sales did not. Anheuser-Busch negotiated such a deal in the late 1960s, under the guidance of Ivy League management scientists.[67] Other consultants with OR/MS experience, including a former head of media and marketing research at McCann-Erickson, accelerated the restructuring of agency compensation in the 1970s.[68]

Though he was ahead of the curve, Harper was not alone in his conviction that social scientists and mathematicians could use information technologies to tether advertising to sales outcomes. Many agency executives saw accountability as not only an affordance of management technique but also an imperative. In 1963 a vice president at Y&R said computers and optimization models "force us to be much more scientific about ways in which we invest our clients' advertising dollars."[69] A year earlier, the president of the MacManus, John & Adams agency told the Association of National Advertisers, "All of us in advertising are going to have to justify and measure our efforts far better than we have been doing. . . . The value-return of every dollar we spend must be justified as best we are able—and with all the scientific assistance we can command—against the supreme criteria of today's industrial dollar."[70]

Increased emphasis on marketing among major corporations, and agencies' efforts to provide comprehensive marketing services, brought advertising, media buying, and market research into tighter integration with industrial operations. Agencies found that in dealing with clients, "top management is paying more and more attention to the advertising function. The result is a growing demand for a more efficient and higher professional approach to every facet of the advertising and marketing process."[71] Economic conditions accentuated these pressures. Pointing to a "profit squeeze" in US business in the 1960s, the president of Booz, Allen & Hamilton warned agencies that advertising spending would decline in proportion to sales. They could expect to feel the pinch as clients demanded more careful justification of plans, more convincing evidence of results, and, overall, "a far more rigorous and quizzical management environment than in the past."[72] Not letting a crisis go to waste, Y&R put an optimistic spin on the situation. In a 1961 ad in the *Wall Street Journal*, the agency welcomed this profit squeeze, which demanded a "cold and calculating eye" and put greater emphasis on research and planning, thereby delivering "a better return on advertising investment."[73] That became precisely the point of agencies' technocratic pitch in the years of intensive computerization that followed.

SHIFTING THE OBJECTIVE FUNCTION

The thrust toward management technique at advertising agencies was largely about reorienting decision-making processes around marketing objectives. Media selection, in particular, was to be judged against metrics that cut closer to companies' profit goals. As a vice president of media at the Cunningham & Walsh agency explained in October 1961, "Media and marketing are synonymous. An advertising medium is a market place, a distribution method.

And with more scientific data to be available more quickly in the future, with statistical relationships determinable by the push of a button, there will result a hastening of the marketing maturity which mediamen should achieve."[74] Less than a month later, BBDO debuted "its 'major breakthrough' in scientific selection of media via computers."[75] By using linear programming to optimize a formally stated objective function, the agency made the integration of media and marketing into an explicit design choice. This apparent breakthrough provoked the industry to openly discuss the purpose of media selection and to materialize commitments to accountability.

BBDO promoted its innovation with Madison Avenue's signature force. The agency took out a full-page ad in the *Wall Street Journal* to boast that it was making more efficient, and sometimes unintuitive, choices by computer-analyzing a "gigantic bank of facts" (figure 8.1). The ad drove home its message in massive font: "Linear Programing showed one BBDO client how to get $1.67 worth of effective advertising for every dollar in his budget."[76] The agency promised to improve ROI by more precisely distinguishing valuable audiences and advertising placements from less desirable ones. By assigning weights or values to different consumer groups, BBDO claimed it could maximize the delivery of not just prospective customers "but specifically prospects most likely to purchase."[77]

Some observers were not satisfied that optimizing advertising exposure among even the most desirable audience segments actually approached the objectives marketers truly cared about. At a demonstration of BBDO's linear programming model, a researcher from Price Waterhouse complained, "This discussion has not once touched on the effect of advertising on sales or profits."[78] A management scientist from Benton & Bowles made a similar observation: "If you don't have some criterion of return, either in terms of profit or sales, then you are maximizing something which is a function of a series of subjective judgments."[79] Critics alleged that BBDO's technique provided cover for a "circular game" whereby media planners could adjust the weights assigned to customer groups and media vehicles until the computer system produced the same selections as conventional methods.[80] But proponents defended BBDO's method as a step toward optimizing marketplace outcomes. The director of applied sciences at CEIR, a data processing firm that helped BBDO build its computer model, explained, "While our criterion (or objective) function is related to [advertising] exposures, it is modified in such a way that it is probably related to advertising effect on the desired

THE WALL STREET JOURNAL, MONDAY, OCTOBER 8, 1962

Linear Programing showed one BBDO client how to get $1.67 worth of effective advertising for every dollar in his budget.

More prospects
A second BBDO client was trying to choose a TV program. Media analysis using LP showed that even though one show had higher ratings, another would reach three times more prospects for the money!

Increased profits
LP showed another BBDO client how to increase profits without increasing the advertising budget. The secret? Concentrate advertising weight where profit from sales would be highest. Nearly an impossible job without LP.

Greater efficiency
LP helped show another client how to increase his ad efficiency 43%—and still maintain the media buys his company is "locked into" because of a complex dealer problem.

Fresh prospects
One of our major clients discovered new media combinations not thought feasible before...so their advertising will reach new, fresh prospects this year.

LP details
LP has been operational at BBDO since April. Part of its magic comes from a computer. But most is from human judgments, skills and facts. A gigantic bank of facts places BBDO at least two years ahead of any other agency attempting to develop a similar program of media analysis. For details, contact Tom Dillon, General Manager, BBDO, 383 Madison Avenue, New York, N.Y.

FIGURE 8.1
BBDO advertised its linear programming system with the promise of better ROI. (*Source: Wall Street Journal*, October 8, 1962, 13; reprinted with permission from BBDO)

audience. 'Probably' is necessary in this statement due to our present inability to measure accurately some of the important variables."[81]

Despite the caveats, mathematical programming looked like progress. Later media selection models tried to move the optimized value even nearer to marketing outcomes. The MEDIAC model, designed by John Little and Leonard Lodish and sold as a computer model utility to agencies, set a measure of market response as its objective function. Little and Lodish declared that their model stood out from others specifically because "the advertising effectiveness process is treated in greater detail."[82] Agencies made similar claims about their own systems, and not just as theater for the trade press. A document detailing the procedures for using JWT's media planning system declared, it is "obvious . . . that new research data, combined with the computer, has increased the capability for evaluating media in terms directly related to marketing problems and to measure more exactly each medium's contribution to the plan."[83]

Nevertheless, analytical techniques could only go as far as measurement capabilities allowed. Behavioral data on media usage and product purchasing were what Thomas Hughes might have called a "reverse salient," a weak component that restricts a system's potential and is thus targeted for development.[84] "We know absolutely nothing about the effect of an advertising message," the manager of Y&R's Media Department admitted. "Measuring impact will have to be the next frontier of media research."[85] Applications of computers and mathematical models accentuated the need for that information. With widespread publicity about automated and optimal decision making, the same Y&R manager explained that "all of us will have to . . . more closely pinpoint the values which advertisers will receive for the advertising dollars they invest."[86] As an article in *Sponsor* magazine put it, "the harsh light emitted by computers" exposed the deficiencies of available data: "computers are forcing media departments to take a long hard look at the data that is being fed into them."[87] Recognizing that "nonsense" data inputs could only lead to "nonsense" computer outputs, an executive from Leo Burnett stressed the need for accurate and personal information. "To be really useful in handling complex media problems, the machines need reliable data on the effectiveness of actual exposure of advertising to real individuals in real households."[88]

Agency researchers, media directors, and management scientists all claimed that computers and mathematical models both demanded better data and stimulated their production.[89] Little called mathematical models "a stone in the shoe for better data," prodding management to measure all phenomena relevant to whatever relationships it wanted to explain, predict, or control. "The model forces explicit consideration of every factor it contains and so pinpoints data needs," he explained in a widely cited paper.[90] Models were like data wish lists, specifying what elements had to be accounted for and what measurements were required to act rationally within the defined reality. Of course, each model was itself a collection of strategic choices about who and what mattered. And it was not computers and models per se that demanded data but rather the imperative to exploit (or at least appear to exploit) the perceived capacities of these expensive intellectual technologies. Still, the practical consequences were much the same.

Researchers who responded to these calls for data recognized advertising optimization as an intersection of management and behavioral sciences.[91] Marketing models developed by the former required "more sophisticated inputs" from the latter, meaning more behavioral data. As a marketing professor from Columbia University suggested in 1970, "A variety of operations research and management techniques lie fallow because of the lack of behavioral input."[92] By then, researchers at Stanford University, with funding from the Ford Foundation and the American Association of Advertising Agencies, were engaged in a "continuing program" of empirical measurement "to fill this void in behavioral information for advertising media models."[93] In particular, they were collecting data on the effects of advertising repetition for use in the MEDIAC online computer system. They used a mobile laboratory provided by the Foote, Cone & Belding agency to measure viewers' responses to repeated TV ad exposure. Interestingly, to distract participants from the true purpose of the experiment, these researchers claimed to be gauging reactions to a hypothetical home shopping system accessed via interactive cable TV.[94] As shown in the last chapter, shoppable electronic media were at the beginning of a winding march toward online buy-now buttons, which are, arguably, the clearest signal of attribution advertisers have. Fifty years later, most Americans not only use devices that let them close the loop between advertising and sales but also carry a digital marketplace, outfitted with behavioral sensors, everywhere they go. In the meantime, market researchers positioned themselves to see more links in the chain of advertising-sales causation by keeping an eye on what shoppers put in their carts.

FROM INSTRUMENTED MARKETS TO ADAPTIVE CONTROL SYSTEMS

Attribution is, essentially, an evidential paradigm that integrates observations of media usage and shopping behavior. The introduction of new technologies into either sphere creates opportunities to produce new data and restructure the terms of advertising deals, inching closer to the integration of media and marketplace measures. The pressure to account for attribution tends to come from advertisers and agencies, but the weight sometimes falls on attention merchants. This pressure is absorbed with ambivalence. Media companies are not eager to make their advertising revenue conditional on sales outcomes. Broadcasters and publishers can reasonably deny any responsibility for the efficacy of advertisers' campaigns; their job is to assemble valuable consumers, and they want to be paid for the whole audience head count. That said, media firms have sought advantage by flaunting their alleged power to influence consumers. For example, WSM Nashville, home of the *Grand Ole Opry*, courted sponsors in 1945 by reimagining its call sign as "We Sell Merchandise."[95] And a 1946 promotion for NBC spot radio insisted that the network's stations "have a consistent habit of getting advertisers' products on the shopping lists of the buyingest people in the most moneyed markets."[96] These sorts of statements are especially pronounced when companies are trying to build or protect a medium's reputation as a reliable marketing technology.

Media companies have supported their claims with social-scientific evidence. In the early 1950s NBC financed a handful of studies on the sales effectiveness of television. Soon, organizations such as the newly created Television Bureau of Advertising (TvB) were crowing about TV's marketing power. At the 1955 convention of the National Association of Radio and TV Broadcasters, the TvB publicized a few favorable findings. For example, in a study conducted by Ernest Dichter's Institute for Research in Mass Motivations, grocery shoppers in five cities were asked to attribute the items they were buying to the "medium [that] influenced their purchase of the individual products." The participants sorted their items into bins representing newspapers, magazines, radio, or television. Shoppers reportedly attributed 54 percent of the items they bought to TV. Whatever its validity, the act of putting purchases into different media bins was a stark way of visualizing attribution.[97]

As manufacturers and merchants implemented universal product codes and optical scanning systems, advertisers looked to observe grocery aisles and cash registers in new ways. Management scientists saw enormous potential for in-store bar-code scanning to generate valuable marketing data and measure advertising effects. Electronic tracking of retail transactions provided a critical infrastructure for conducting controlled experiments to produce attribution claims. Together with the discriminating affordances of addressable cable television, scanning at the point of sale gave marketers a living laboratory to study advertising.[98]

By the 1970s, packaged goods companies were mining intelligence from "instrumented markets." In these cities, all the major grocery stores were equipped with optical scanners, and cable systems were capable of "splitting" subscribers into treatment and control groups and then distributing different advertisements to households within the same neighborhood. Researchers measured purchases using interviews and pantry audits, and stores in these test markets eventually introduced identification cards to recognize consumers and automatically document what they bought. By combining household-level measures of advertising exposure with observations of purchasing behavior, "a remarkably complete picture of the shopper's marketing environment is possible."[99] Within this naturalistic experimental setting, researchers could conclude that differences in relevant consumption behaviors across treatment and control groups were "attributable to the controlled television exposure."[100] Celebrating the combination of in-store scanning and control over advertising delivery, the president of the ARF wrote, "*These technologies have brought the power of the microscope to advertising research, along with enhanced experimental design capabilities, and increased productivity and analytic power.*"[101]

This was a familiar area of interest for management scientists like John Little. The isolation of advertising effects was precisely what he and his colleagues had been attacking with OR/MS methods in the 1950s. Throughout the 1960s Little pursued this agenda by devising "adaptive control systems in marketing." Basically, he proposed a recursive process of building a model to represent a marketing process, continuously experimenting with advertising policies to generate data about consumer response, and using what was learned from those experiments to optimize the model and adapt advertising strategy. Describing how this approach should be applied to the problem of setting an advertising budget, Little and Russell Ackoff emphasized the importance of responding dynamically to marketplace feedback: "what we really want is automatic controls, the purpose of such controls being to keep the spending in each market as close as possible to the point of maximum profitability for the company."[102] They wanted something like programmatic advertising. The manager of marketing research at Coca-Cola, Malcolm McNiven, described Little's designs as possibly "the most fruitful approach for measuring effectiveness of advertising."[103] Little and his collaborators commercialized their products through a firm called Management Decisions Systems (MDS). MDS sold advertising models, analyzed consumer behavior for marketers, and built management information and analytics software, including a program called EXPRESS, which MDS sold to Oracle for $100 million.[104] Retail tracking technologies presented another fertile area for the company.

Marketplace data represent "contact points with reality," Little said. "Whenever new measurement technologies appear, they create special opportunities for learning how the world works."[105] Writing with his student and MDS staffer Peter M. Guadagni, Little likened the improvement in measurement afforded by in-store optical scanning to Galileo's telescope.[106] These systems seemed capable of delivering a more powerful, dynamic view of consumption. Looking ahead from 1979, Little projected that within five to ten years the amount of transaction-generated data and the computer power available to analyze those data would increase by an order of magnitude.[107] The speed and scope of automated data collection would then permit management to pinpoint how sales are responding to advertising moment by moment, rather than looking back at recent sales. Retail tracking also facilitated the production of "longitudinal customer histories."[108] "The chief advantages of the scanner panels lie in their micro detail and competitive completeness," Guadagni and Little wrote. "People, not markets, respond to the actions of the retailers and manufactures."[109] By 1987, Procter & Gamble's manager of information services seemed to confirm Little's earlier projection: "Over the past five years, new ways of reading consumer behavior have emerged, and most are electronic; that will continue. That provides people who study consumer behavior an immense, rich new database."[110]

In 1985 MDS merged with Information Resources Inc. (IRI), whose BehaviorScan service was an industry leader in furnishing instrumented markets for companies to experiment with advertising and new products. Little joined the IRI board, and one of his students, Magid Abraham, later become IRI's president. (Abraham went on to found comScore, an online audience measurement firm.) IRI was doing so well that in 1987 Dun & Bradstreet tried to buy the company for $570 million.[111]

These integrations of media and shopping data came closest to determining advertising response. They also raised concerns about privacy.[112] In his exposé on database marketing, *The Naked Consumer*, Erik Larson writes, "The attraction of scanner technology is powerful. It lets companies observe market phenomena they previously couldn't have seen. It advanced the marketers far along in their century-long drive to turn their art into a science."[113] Many advertising professionals were excited rather than concerned about exposing more of consumers' lives to detailed study. They saw an opportunity to finally access a supply of behavioral data worthy of their adaptive control systems and the conceit of optimization. As one agency executive said about retail scanners, "A device conceived for operational reasons to control inventory and to speed up and improve the pricing accuracy at the checkout will turn out to feed the marketing data base of the 1990s and beyond."[114]

FALLING OUT OF AND INTO FASHION

These rapid and granular measures of consumer response provided the data that futuristic media departments had been clamoring for since the 1960s. But, despite all the ballyhoo about technoscientific breakthroughs, by the time these data were widely available, the optimum-seeking models described throughout this book seemed like artifacts of the inflated expectations surrounding OR—a legacy of overpromising and underdelivering. During the 1980s and early 1990s, routine practice in media departments settled around software tools that were far less complex than the models designed by leading operations researchers. The latter aimed at the objectives and distinctions marketers desired, but their potential to extract new value was circumscribed by organizational conditions such as how attention merchants packaged their inventory. A management scientist at BBDO described this in 1981 as a "classic implementation problem," wherein "modeling specialists" had not adequately considered the workaday needs of media planners.[115] Furthermore, while the independent media buying agencies that rose to prominence in the 1970s and 1980s marketed their services *in part* around calculative expertise, they emphasized much more forcefully their ability to negotiate low prices. Clout and personal relationships figured into this style of buying more than strict rationality,[116] due in part to the concentrated power of media capital in the United States. The optimization models hyped in the 1960s and 1970s found wider use in Europe (albeit with American money), where more media buying and selling were done according to a rate card and where the political economy implied different bargaining relations.

But in the late 1990s, at another moment when technological, political, and industrial changes invited bold claims about progress, optimization models once again enchanted the US advertising industry. Major advertisers, including the formidable Procter & Gamble, which had a $1 billion TV budget, openly demanded that their media agencies use optimization models, and those agencies focused more energy (at least in their public performances) on shifting the objective function in their planning toward closer approximations of ROI. Audience fragmentation was intensifying; this complicated existing business practices but also implied lucrative possibilities to discriminate among consumer populations. These challenges and opportunities opened a problem space that renewed the salience of optimization and accountability. As the director of North American media services at JWT put it in 1998, "Why we're all so concerned about optimizers [and] return on investment is we know we're losing the mass media and we have to change and take a look at tomorrow, and those are the tools that help us look at tomorrow."[117]

The future was up for grabs again. "The next generation of optimizers will seek to match viewing information with purchase data," an article in *Fortune* explained. "Eventually, buyers would like to determine with some exactitude just how their spending on commercials translates into sales."[118] Leading agencies and media buying services continued investing in research to find out "if and why some kinds of TV exposures produce greater consumer response—and to quantify these differences to make their optimizers smarter." As one expert put it, these studies "will then take us closer to optimizing the Big Kahuna—product sales."[119]

The two main strands of OR/MS in advertising—optimizing media selection and accounting for advertising effectiveness—were recombined. The advertising industry was becoming completely interwoven with direct marketing, electronics and software companies, and proliferating research and information management services. It rode this wave into the twenty-first century, propelled by the warm winds of state-sanctioned internet commercialization, financial speculation, and an unregulated data buffet. Over the next two decades, advertising money flowed toward digital media that promised to produce more complete and perfect records of advertising events and consumer behaviors.[120] And as more and more shopping moved online, marketers believed that the entire chain of causation between advertising and sales could be captured within a digital evidential paradigm. Digital ad servers and social media platforms are evidence machines, combining unprecedented accountability with the other affordances described in previous chapters. Today's adtech materializes the spirit of using adaptive control systems to measure and maximize ROI.

While the data and analytics supporting attribution remain riddled with problems, the advertising industry has, in practice, implemented an evidential paradigm that integrates records of media and marketplace activity. Companies like Google, Meta, TikTok, and Amazon market their advertising businesses around their fortressed access to observations of consumer behavior, their machine-learning capacities (which benefit from huge training data sets), and their power to orchestrate activities and configure choice architectures. Recently, they have been angling to conduct more consumer purchasing directly on their platforms, in the same environments where they serve ads. These companies vary in how much they prioritize and succeed at this effort, but overall, they come as close as anyone to fulfilling the dream of enclosing consumers within management science machines.

Platforms are the envy and the economic gravity of the advertising world. Their size and their integrations with partners that depend on them (e.g., publishers, merchants, app makers) help them account for a massive inventory of variables related to the expected value of consumers and the apparent success of advertising events in delivering desired outcomes. Their attribution measures are recycled back into automated decision systems, ostensibly optimizing ad spending and distribution to achieve programmed objectives. Platforms can also break up micromoments and microbehaviors into salable advertising opportunities. Their abilities to account translate into abilities to bill.

A great irony, however, is that as these marketing intermediaries built the accountable systems advertisers had been dreaming of for decades, those advertisers and their agencies lost control. For starters, the documentation of audience attention and behavior has shifted away from ratings produced by third-party measurement firms and toward server logs generated by whichever party serves an ad. The promise here is a full census measure of advertisement distribution; however, it also means that those accounts are produced through a sort of internal auditing by adtech intermediaries that are direct parties to these transactions, rather than an ostensibly neutral arbiter.[121] Given the scale and opacity of digital advertising supply chains, advertisers often have no idea where their ads appear. The business of fabricating evidence of audience attention or behavior, though technically capable of accounting for more events than ever before, has become less transparent to advertisers and publishers. The authoritative production and circulation of this information are critical levers of power in these industries, and adtech intermediaries have claimed this power as their own—even though they seem incapable of knowing what happens with all the data they collect.[122]

Furthermore, the complexity of models and analytical techniques and the relations of data generation and control prevent advertisers and agencies from confidently adjudicating attribution claims. The drive for accountability—where advertisers can monitor, measure, and record all consumer behaviors in response to advertising—has empowered the intermediaries that appear most capable of untangling the knot of attribution. Advertisers asked technical experts to get to the bottom of advertising effectiveness, and eventually the experts dug so deep that they disappeared from sight. Optimization models designed by management scientists in the 1960s gave accounts of *process*, accounts that confirmed rational calculations and commitments to objective values. Impressing with their technique, they legitimated uncertain decisions whose desired outcomes were beyond measurement. Today's management science machines give accounts of *outcomes*—in fact, they are constantly generating accounts (of patterns, correlations, deviations, value). But the processes by which they engineer those outcomes may be undisclosed or, especially with machine learning, impossible to explain. The objective function and the focus of measurement have shifted toward marketing outcomes, as advertisers hoped for, but authority and trust shifted too. The search for exhaustive detail has yielded mystification and made it a source of market power.

It is worth noting, as well, that ad fraud, totaling tens of billions of dollars a year, is the flip side of a business built around the production of accounts. Its whole basis is fabricating information to imitate legitimate evidence of attention, behavior, or value. Fraud has thrived in a business culture that fetishizes documentary records and metrics and the success stories they help tell.

CONCLUSION

In 1964 Charles K. Ramond, the ARF's technical director (and formerly a researcher at DuPont), wrote a flattering prediction about how a scientific attitude would turn advertising into a prestigious modern profession. He wrapped his forecast in a fanciful conceit—presenting it as a letter from the year 1999, obtained via time machine, from historians who explained "how the advertising business was revolutionized by the simple expedient of measuring the profitability of its expenditures." Looking back from the edge of a millennium, these imagined historians wrote, "Advertising became a measurable contributor to the business firm late in the twentieth century. Until that time, virtually no manufacturer had any real information about the economic effects of the money he spent to advertise." The industry crawled out of the dark ages when "several major U.S. companies—among them Du Pont, Ford, General Electric, and Scott Paper—began to investigate systematically the relationships between advertising expenditures and the sales they caused." Their successors committed to gathering "continuous experimental evidence," and eventually managers possessed "rapid knowledge of the consequences of their marketing decisions (artfully presented on 3-D color television)." By the end of the century, "documentation of economic effects" precipitated the "gradual disappearance of objectionable advertising"—a reasonable encapsulation of the "relevance" argument mobilized today to defend behavioral targeting. It's an imaginative little scene, and it appears prescient. But like most works in this genre, it was as much a suggestion as a forecast.[123]

Today, sci-fi contrivances are apparently unnecessary. Enthusiasts say the future is now. Datafication, analytics, and administrative capacities let marketers account for the value of specific consumers and the productivity of advertising efforts. By 2017, an executive at IBM's artificial intelligence (AI)–focused Watson Advertising declared demographic targeting "a kind of historical artifact" reflecting the limits of what he described as paper-based means of information and transaction processing. Marketers and media companies no longer needed to use demographic categories to simplify human behavior and identity and to package people into standardized bundles, he claimed, "now that technology and infrastructure allow us to . . . really understand how the individual functions—what they buy, where they go, what their interests are, what media they're consuming." With AI and big data, "now we can try to at least calculate the value of the individual and what that represents for a marketer at a moment in time and at a place."[124]

Accountability bookends advertising's calculative evolution across the second half of the twentieth century. The dream of determining the sales effects of advertising was a primary entry point for OR/MS in marketing. Advertising has always tried to manage demand; OR/MS spearheaded technocratic efforts to rationalize the sphere of consumption and absorb it into the scope of management, optimizing consumer behavior for private profit. The proliferation of measurement instruments—from the digitization of retail environments through the popular adoption of mobile devices and the expanding internet of everything—has been shadowed or even spurred by aspirations to control the circulation of commodities. Fantasies of perfect accountability run through many designs to make media into extensions of the marketing system.

This chapter has shown how those fantasies found material, intellectual, and organizational force. The attitude and techniques of cost accounting inflected the OR/MS approach to advertising, favoring the seemingly concrete behavioral metrics of direct marketing over the psychological or cultural effects of branding. OR/MS's expertise excited advertisers' desires to measure and predict the effectiveness of their advertising policies and to appraise customers' value more exactly. Determining ROI was the fundamental mandate for the consulting work operations researchers sold to advertisers. For advertising agencies, which were responsible for deciding how to allocate advertisers' money, these pressures toward rationalization and efficiency manifested in the work of aligning media selection with clients' marketing objectives. Accountability became a buzzword in advertising management in the early 1960s. It stood for the scientific and data-driven stewardship of clients' ad budgets, squeezing the most value out of every dollar spent. The computer-powered optimization models designed to automate media selection materialized this spirit of accountability and ROI-centered management.

The shortcomings of agencies' initial efforts accentuated the urgency of producing behavioral data—precise and persistent records of what kinds of media people consumed and what goods and services they bought. Investments in complex models and expensive computers, and the urge to exploit their capabilities, amplified a datafication imperative in advertising. The desire to incorporate finer and more comprehensive behavioral measures into media buying and selling was in tension with computational limits and pressures to standardize data. Simple models of marketing processes sacrificed realism, but they were easier to operationalize and understand. How to accommodate more of reality into the framework of routine planning, evaluation, and action was a challenge for management information systems and the culture and politics of organizations. It was also a point of friction between advertising buyers and sellers—negotiating how an expanding inventory of consumer behaviors and attributes could be packaged into salable evidence of audience attention or action. Advertising transactions needed to accommodate differently datafied consumers and events.

These challenges were not resolved until well after OR/MS had faded into the background in advertising. But the affordances that operations researchers and agency media departments constructed around computers and optimization techniques helped break open a particular horizon where surveillance, prediction, and rational management seemed desirable and necessary. It also helped empower and reproduce a class of techno-scientific experts now referred to as "math men." The OR/MS inflection in advertising and media, for all its failures and false starts, was an engine driving toward more data creation and use and toward a corporate culture where behavioral evidence became an authoritative and expected element of optimal decision making. Management technique—combining computerization, operations research, and mathematical modeling—suggested the potential to refine accounting abilities in advertising. The commitment to optimization was not just data driven; at this pivotal moment in advertising's history, it was dramatically data *driving*.

CONCLUSION

[Technology] is above all a special kind of power to make the world of dreams real. Technologies are embodied hopes, devices to implement beliefs about how the world could be made different. Technologies are social dreams and fairy tales in action.

—Carolyn Marvin

"Technology" is a new suit of clothes for the emperor.

—Dallas Smythe

The concerns addressed in this book are sometimes dismissed as frivolous. *It's just advertising—what's the big deal?* But for many people in the thick of the industry's calculative evolution, the stakes could not have been higher. As Jackson Lears explains, "The effort to ally advertising with the American Way of Life, begun in the dark days of the Great Depression, came to full fruition in the ideologically charged atmosphere of the Cold War."[1] That atmosphere precipitated some particular attachments. Cultural historians have concluded perceptively that advertisements showcase social desires— distilled images of what some people picture as the good life.[2] Business and technological histories support a similar conclusion: industrial discourses about advertising distill corporate desires. As ministers to the business world, advertising professionals cultivate and crystallize not only the fantasies of consumers but also the fantasies of capital. Madison Avenue, and schools of management as well, reflect and refract political-economic cultures. Looking at technoscience in advertising and marketing provides a unique perspective on some of the ways capital dreams of digital futures.

Let's return to where we began this book, at the 1961 convention of the American Marketing Association. There, the vice president for public relations at AT&T painted a vivid picture of the road ahead. The United States faced existential threats, he warned, "perils almost without precedent," from the spread of communism, independence movements against colonialism, and declining faith in profit as a sacred moral doctrine. He assigned technologically sophisticated marketing a major role in the defense of US capitalism, holding a transistor between his fingers to symbolize an information revolution (and to exalt Bell Telephone Laboratories). With the explosion in the capacity to process and transport data, marketing was poised to coordinate more tightly with all levels of corporate management. As he saw it, "a potential is rapidly being developed" to generate and circulate moment-to-moment information about sales orders, inventory, and other business factors. This feedback could be applied, almost instantly as each transaction took place, to improve control over production, distribution, and marketing. He described, in effect, a cybernetic system powering the capitalist economy. Crucially, though, he insisted that government should not supervise or plan economic action. This corporate control system would simply accelerate and optimize free-market processes. It was the marketing profession, then, that should carry out the duties of being "the leader in the era of automation" and "maintaining our American way of life." Like the neoliberal economics of Milton Friedman (whom he quotes in the speech), this technocratic vision disguised a muscular political project. "Above all," the AT&T executive urged, "marketing must demonstrate its vital role in our free society, not only through the satisfaction of the needs of consumers but through the creation of profit." It "must constantly demonstrate to the nation and the world at large that . . . a free market is one of the principal elements of that [free] society." He concluded with an apparent joke: "After all, who ever heard of a Vice President—Marketing in Moscow."[3]

For those fighting the Cold War on economic grounds, and especially those who saw consumer culture as a bulwark against radical class politics and crises of overproduction, the development of a weapons-grade science of consumption engineering became an active front in rationalizing US capitalism. Midcentury marketing began to mobilize "math men," computers, and a powerful ideology of optimization. Across the next five decades, many tactics, mind-sets, and metrics from management science, direct marketing, and the industrial art of data processing were brought to the center of the advertising system and projected onto the problems and opportunities facing corporate America. Advertisers, agencies, consultants, and information and computer service vendors preoccupied themselves with calculating efficient ways to achieve marketing objectives. This was hardly a new ambition, but by framing it as something that could be solved with mathematical models and high-speed computers, a loose community of technical experts helped legitimize particular priorities, modes of reasoning, and hierarchies of knowledge and authority that are entrenched today at the heart of digital advertising. These approaches reformulated advertising operations in terms of probabilistic and algorithmic decision making. Advertising distribution became a problem of optimizing an "objective function" representing a measurable goal such as profits, sales, or access to valuable audiences. This required planners, decision makers, and the workers they managed to formalize and account for the relationships representing marketing processes, including consumers' responses to advertising. What they built, in fits and starts, was adtech.

Modeling and automation in advertising involved concerted efforts to program decision processes and to assemble calculable consumers so that both could be acted on by evolving forms of machine intelligence, with all the speed and precision that seemed to imply. These maneuvers gave urgency to datafication efforts to incorporate more of reality into the frame of accounting and rationality. Predictive and evaluative models required explicit, quantitative statements about the value of advertising opportunities and consumer populations and the probabilities of certain outcomes. The dream at the end of the rainbow was that management could predict and even manipulate consumer behavior, much as operations researchers manipulated factors of production and distribution. Marketers aspired to bring consumption under managerial control—to document and materialize consumers, audiences, and media channels such that they could be governed by management science machines. Although marketers have not found, and may never find, the gold where that rainbow meets the road, their search has motivated widespread efforts to count, classify, and intervene into more of daily life. And it might not matter if the rainbow is a mirage. Adtech works by enacting and seizing value from a crowded universe of potentialities—by trying to make bets that will pay off more often and at better margins. Whether or not those bets lead to behavioral manipulation, either directly or indirectly, this system of panoptic sorting affects allocations of power, access, and possibility in society.

The calculative evolution that helped usher in surveillance-based advertising is thus not a sharp departure from the supposedly discarded logic of predigital advertising but rather a particular inflection of that logic via management technique. It represents a rationalization, pushing toward the logical conclusions of selling the American people. In an important sense, twenty-first-century adtech recombines advertising's calculative evolution with its better-known creative revolution. The creative revolution venerated individuality and difference as the spiritual center for a personalized style of consumerism. But the range of possibility opened up by regarding consumers as unique and spontaneous was hard to reconcile with business processes built around routines, standards, and convenient but imprecise categories and valuation methods. The commercial shaping of online and mobile media and their infrastructures of surveillance and big-data analytics promised a breakthrough—an exacting ability to manage and capitalize on the dynamism of a flexible mode of consumption. What Mark Andrejevic calls the "digital enclosure" materialized long-standing efforts to accommodate the fluid and personal details of daily consumer life within expanding and adaptable systems of measurement, calculation, management, and optimization.[4] Now-dominant parts of the advertising industry leveraged big and small computers, databases, and data sciences to bureaucratize the counterbureaucracy—to make the new consumerism accountable and to market those accounting practices and the data they produced.

As much as creative production and other areas of management thinking turned away from bureaucracy and control, the already financialized and technoscientific wing of advertising never gave up on its dreams of optimization. In adtech, as elsewhere, anomalies, nonconformity, and difference are all made productive for speculation and arbitrage. Particularities of lifestyle and individuality, which overflowed the accounting and transactional capacities of mid-twentieth-century advertising, have been incorporated into systems designed to discern consumer propensities and thereby engineer new actionable value claims whenever attention or behavior is packaged and sold. Transactions that can exploit these microscopic opportunities have been constructed around addressable communications that discriminate among message recipients and automation technologies, including programmatic auction platforms, that accelerate flows of information and commerce. Furthermore, with ubiquitous connectivity to shoppable media, it is possible to see, sometimes in an instant, whether a bet pays off. Overall, advertising's calculative evolution has been punctuated by repeated efforts to transform moments of existence in a consumer society, with all their variety and spontaneity, into differentiated investments scored by probabilities and profit projections.

Looking at these developments through the lens of affordances, or perceived potential, shows how dreams of optimization have helped define information technologies and privilege certain uses and users. Affordances are about power and vision. Their specter is always present in audience commodification, which is essentially the creation of market relations around a potent abstraction. That abstraction, whether we call it audience, attention, or behavior, represents potential value backed by some documentary evidence, and it can be redefined almost endlessly to channel money, data, and advantage to different participants or in different proportions. This flexibility, and the political process of interpreting technologies according to their apparent capacities to alleviate certain crises and capitalize on certain opportunities, is key to understanding adtech. Adtech emerged through strategic and sometimes contradictory investments in an array of artifacts, techniques, and dispositions of optimum seeking. Optimization is, almost by definition, never final, and signs of either success or failure can fuel aspirations of progress—to know more, to predict more accurately, to evaluate more exactly, and to do it all faster. Optimization is a claim about a future state of the world where some latent or hidden potential is productively exploited, usually thanks to better knowledge or logistics. It is a promise of *more perfect* that is constantly recycled and deferred and almost always seductive.

Adtech's code, data, and day-to-day practices animate a tangle of meandering but deep-rooted ideas about what information technology could do and could be. In *The Closed World*, Paul Edwards asks how computerized control systems came to seem like a technological possibility in Cold War America; how particular quantitative forms of management challenged older hierarchies of expertise, authority, and judgment; and how fantasies of control remained credible, and even gathered influence, in spite of evident failures.[5] This book has pursued a similar set of questions, focusing on commercial adaptations of computational and optimum-seeking management sciences. Like Edwards, I find that optimization in advertising drew strength not just from the undeniable force of mathematical proof but rather from the stories, narratives, and discourses that provided resources for well-positioned actors to assert competency, to deflect blame or take credit, and to legitimize their visions for the future—a future brought within reach by the perceived potential of new technologies and techniques.

Today's adtech companies seized opportunities to intermediate business and to assert infrastructural power, promising the same suite of affordances that some in the advertising industry had been trying to foist onto information technologies for decades. By the time policymakers began conspicuously regulating the internet in the interest of capital accumulation, advertisers and marketing intermediaries had developed a knack for linking new media to an ideology of optimization. Through the spread of networked devices packed with sensors and trackers, governments, corporations, and individual consumers helped subsidize an extensive infrastructure for behavioral measurement and dynamic intervention. As this bedrock for adaptive control systems was cobbled together for a variety of purposes, advertising specialists exercised a handy set of discursive tools to promote particular commercial uses. Internet communications and the whole package of servers and databases, interactive applications, and measurement and analytical techniques seemed capable of combining *automation, addressability, shoppability,* and *accountability.* In many ways, digital advertising is defined by exactly those capabilities. The internet's marketing affordances had already been imagined and invested with perceptions of enormous economic potential, while the materials and organization necessary for realizing them were still taking shape.

This book shows how the evolution of marketing and audience commodification has involved repeated attempts to harness technologies that afford precise and rapid discriminatory judgments. Adtech is part of this longer history of computerization in advertising, of ongoing efforts to accommodate more detailed accounts of expected profit and risk within industrial routines. Google, Meta, and other adtech and data brokerage firms have consolidated power by building themselves into computational and logistical utilities; they fabricate and digest information, they make claims that order the world for rational decision making and optimizing, they orchestrate and accelerate transactions, and they link those processes at critical bottlenecks. Platforms that monetize sociality have been especially effective at enclosing countless relationships and activities within calculative environments.

It is clear that discrimination and surveillance have been central to the mutual shaping of information technology and the political economy of media in capitalism. This recognition forces us to deal with digital advertising's pathologies in ways that address a key underlying cause: the corporate will to produce and exploit differences in the probable value of people and the likelihood of their responding to strategic influence. What does an organized movement against adtech and surveillance capitalism look like if we understand them not as aberrations but as the automation and extension of core capitalist tendencies via management technique?

First of all, we can commit to radical change. The focus on optimization is helpful because, in a way, optimization is a radical process, in that it reaches toward the fundamental logic of a system. Despite pretensions of apolitical efficiency, optimization begins by declaring what's at stake—who and what matters, what state of the world is desirable. Marketing models are mathematical statements of purpose and priority. What these models account for, and what they externalize, tells us how the model makers and users see the world. Advertising optimization could be aimed at a wide range of objectives, but, for the most part, the specified goal has been to maximize some measure of private value. This matters a great deal in capitalist societies that have delegated the responsibility for financing communication resources and cultural production to marketers. Our examination of optimization models reveals that advertisers are not explicitly investing in things like democracy, diversity, justice, honesty, or most other values that various social traditions consider important to public and private life. To the extent that the integrity of media sources and the quality of media content are considered at all, they are evaluated through advertising priorities: do they organize the apparent attention of useful consumer populations, and will they help or hurt the persuasiveness of commercial and political messages? Programmatic advertising virtually eliminates any responsibility for supporting democratic media, as marketers finance the arrangement of advertising events by almost any means, irrespective of the surrounding content. To put it bluntly, advertisers don't measure return on investment in terms of quality of life. They are trying, through adtech intermediaries, to optimize their campaigns to spend no more than necessary to achieve business objectives. This should tell us that their commitment to the social value of media is at best incidental to the goal of tethering media to marketing. Counting on that set of relations to serve democracy or to produce information, entertainment, and experiences that enhance public rather than private interests is like praying for a positive externality—hoping for an outcome that is not explicitly prioritized.

This all deals a serious blow to the argument that advertising is necessary to sustain democratic media systems. One of the most important conclusions we can draw from zooming in on the inflection point in advertising's calculative evolution—when the accommodation of computers and management science forced implicit values to be articulated—is that advertisers and the specialists that help them pursue their objectives should not be trusted to organize publics, public spheres, and public life. The calculative techniques described in this book reveal, explicitly and formally, what advertisers care about and what they want to accomplish. They tried to refine their ability to optimize private value by gathering more data about consumer behavior, assembling more calculative firepower to appraise profiled individuals and populations, and taking advantage of more invasive and intimate opportunities to target their influence. As more of the technologies that mediate social life have been integrated with marketing and its systems of knowledge and power, the logic of selling the American people has become a design principle that organizes access to a staggering range of experiences. Clearly, though, it is not a democratic design principle.

But other plans are possible. Many promising policies, proposals, and political movements for democratizing media and information are already at hand.[6] I will not describe them in detail here, except to point out that irredeemable problems with advertising—not just adtech—should alleviate any hesitation about pursuing dramatic and imaginative change. What is needed is nothing less than "a new institutional structure of public operation and control" for the diverse range of networked information and communication utilities that serve communities' needs.[7] The United States desperately needs what Sanjay Jolly and Ellen Goodman call "a 'full stack' approach to public media," engaging libraries and schools, for example, as both community spaces and operational hubs for search engines, social media, and local broadcasting.[8] Policies and inclusive political action should focus on more than the governance of platforms, data extraction and use, and algorithmic systems; these are critically important, to be sure, but they sometimes absorb a disproportionate share of discourse and energy and can even be distracting when they promise simplified solutions that ignore the broader ecologies of media, culture, and politics. Radical reform of political campaign financing, broad support and protection for labor organizing, sharp restrictions on the career pathways and patronage that lead to regulatory capture, universal broadband access, intersectional coalitions that link local and planetary struggles for liberation and peace—these are just a few critical building blocks of democratic communications. Nevertheless, passing comprehensive federal laws regulating privacy and personal data, and implementing focused interventions around surveillance advertising and antitrust are urgently needed. The details of such interventions are far from settled, but the political project has to go beyond curbing undeniable abuses or tweaking the dials of privacy and competition within existing market arrangements. It must aim, instead, at reorganizing media institutions and infrastructures around different purposes, priorities, and power relations. As Dan Schiller argues, a sustainable and democratic way forward requires "a far-reaching process of decommodification. . . . Advertising must be eliminated as a funding mechanism for essential services, and replaced by government support, with measures to insulate these services from political meddling."[9]

The specter of any meaningful reform to the system of advertiser support often brings an outpouring of melodrama from the system's defenders. Even calls to limit tracking and profiling, or to consider alternatives to markets and private profit as the respective mechanism and purpose of data governance, are met with fearmongering about the end of the "open web." *The sky is falling,* they say.

The claim that we can't touch advertising for fear of harming the internet must be dismissed. First of all, the complaint that the sky will fall without behavioral advertising ignores the fact that it's already raining meatballs. The commercial web and the walled garden platforms have profound structural problems.[10] These problems are hidden to some extent by the astonishing force of human creativity—the fact that beautiful people can make beautiful things even in a hellscape. But fearless journalism and uplifting, empowering, and entertaining forms of content and mediated experience persist *in spite* of advertising and its dream of organizing existence into risk-rated investment opportunities. The suggestion that the web is too perfect or precious to change betrays an embarrassing lack of imagination and political will.

The claim that reductions in advertising will cause news and investigative reporting to be stashed behind paywalls raises important concerns about inequity. But this claim understates the extent to which the commercial news already reflects a logic of discrimination, centering the needs and tastes of mostly white, middle- and upper-class consumers.[11] As C. Edwin Baker argued decades ago, advertiser support subsidizes content reflecting the interests of audiences who could afford to pay for that content, while underproducing for poor and working-class audiences.[12] It would surely be troubling if only wealthy people could access accurate news, but this debate needs to consider how cultural production and distribution would change if the pressures and incentives of advertising were diminished. In short, inequity is already a problem. We should address it through deliberate and direct action.

Relatedly, Meta and other organized interests argue that restrictions on tracking and targeting will disproportionately affect small businesses. This is a sympathetic claim, but it sidesteps a larger conversation about whether we should use surveillance advertising as a mechanism to deal with failures of competition policy and antitrust enforcement. It is not obvious that protecting entrepreneurs should be a defining purpose of communications media, especially if those businesses cannot afford to acquire customers without using cheap ads that externalize the costs of data extraction and profiling. But if supporting small businesses is a social goal, it ought to be treated directly with policy and regulation, rather than as a trickle-down effect of discrimination-as-efficiency.

Perhaps the most disingenuous aspect of arguments for the status quo is that they seem to assume that any political action affecting advertiser support would not be coupled with active measures to encourage the provision of useful information and communication services. If a bridge was deemed a safety hazard but a necessary conduit for traffic, no serious debate would be limited to the two options of either tolerating the threat of collapse or demolishing the bridge and walking away from the rubble. People want and need news, entertainment, social connection—the whole range of purposes for which humans have adapted capitalist media. We want a bridge to reach those purposes. But an advertiser-supported media environment is, on balance, a hazard—to equity, to justice, to sustainability.

Well-meaning commentators insist that advertising does not necessarily devolve into surveillance and discrimination. While that may be true in principle, historical evidence suggests it is rarely true in practice. Surveillance advertising exists because of ongoing, if uneven, efforts to use media systems to extend and accelerate the circulation of commodities and to produce evidence of attention and behavior as investment opportunities. Attacking the roots of today's problems requires more than just eradicating adtech's obvious offenses, such as shady tracking and profiling, rampant fraud, monopoly power, monetization of hate and deceit, and illegal discrimination. It is certainly good to squeeze out bad actors and choke off the revenue that funds blatant disinformation and hate. But convincing corporations to subsidize brand-safe content for affluent consumers is not a serious solution if we hope to confront inequality, injustice, and consumerism's environmental impacts. Mainstream ad-supported media, prone to distortions and market failures, have long histories of producing subtle but systemic misinformation, framing social issues in ways that legitimize entrenched wealth and power and omit or discredit the concerns of people who are marginal or resistant to an established order.[13] Real progress will require coming to terms with the grave implications of capitalist media and information systems and recognizing that surveillance and discrimination are central to selling the American people not only online but also throughout commercial media. Advertising may have a place in society, but that place cannot be as an organizing institution of social infrastructures.

This book has shown that well-worn arguments about the necessity of advertising do not hold up. Advertising is a dubious system for structuring access to culture and social life. It is beset with contradictions, and the attempts to resolve those contradictions or to make the system more rational have mostly made it worse. The ideology of optimization chases private values, with social values externalized as afterthoughts. "New information technologies will be used by the powerful to increase their power unless somebody makes other plans," Carolyn Marvin warns.[14] Media systems for a democratic, just, and sustainable future will require us to dream and design differently—to make other plans.

NOTES

INTRODUCTION

1. "The Curse of Stereotyped Marketing," *Broadcasting*, June 26, 1961, 48; Albert W. Frey, "Approaches to Determining the Advertising Appropriation," in *Effective Marketing Coordination: Proceedings of the Forty-Fourth National Conference of the American Marketing Association*, ed. George L. Baker Jr. (Chicago: AMA, 1961), 332.

2. Russell H. Colley, "Squeezing the Waste out of Advertising," *Harvard Business Review* 40, no. 5 (1962): 77. On rationalization in US manufacturing and management, see David F. Noble, *America by Design: Science, Technology, and the Rise of Corporate Capitalism* (New York: Alfred A. Knopf, 1977); James R. Beniger, *The Control Revolution: Technological and Economic Origins of the Information Society* (Cambridge, MA: Harvard University Press, 1986).

3. Frey, "Approaches," 327.

4. Frey, 332; "The Curse of Stereotyped Marketing," 48.

5. Frey, 335.

6. Paul N. Edwards, *The Closed World: Computers and the Politics of Discourse in Cold War America* (Cambridge, MA: MIT Press), 114–115.

7. Oscar H. Gandy Jr., *The Panoptic Sort: A Political Economy of Personal Information*, 2nd ed. (New York: Oxford University Press, 2021); Helen Nissenbaum, *Privacy in Context: Technology, Policy, and the Integrity of Social Life* (Stanford, CA: Stanford University Press, 2009); Joseph Turow, *The Daily You: How the New Advertising Industry Is Defining Your Identity and Your Worth* (New Haven, CT: Yale University Press, 2011); Mark Andrejevic, *Automated Media* (New York: Routledge, 2020); Greg Elmer, *Profiling Machines: Mapping the Personal Information Economy* (Cambridge, MA: MIT Press, 2003); Josh Lauer, *Creditworthy: A History of Consumer Surveillance and Financial Identity in America* (New York: Columbia University Press, 2017); Frank Pasquale,

The Black Box Society: The Secret Algorithms that Control Money and Information (Cambridge, MA: Harvard University Press, 2015); Sven Brodmerkel and Nicholas Carah, *Brand Machines, Sensory Media and Calculative Culture* (London: Palgrave Macmillan, 2016); Solon Barocas and Andrew D. Selbst, "Big Data's Disparate Impact," *California Law Review* 104 (2016): 671–732; Marion Fourcade and Kieran Healy, "Seeing Like a Market," *Socio-Economic Review* 15, no. 1 (2017): 9–29; Jake Goldenfein, *Monitoring Laws: Profiling and Identity in the World State* (New York: Cambridge University Press, 2020); Salomé Viljoen, "A Relational Theory of Data Governance," *Yale Law Journal* 131, no. 2 (2021): 573–654.

8. Nick Srnicek, *Platform Capitalism* (Cambridge: Polity, 2017); Robin Mansell and W. Edward Steinmueller, *Advanced Introduction to Platform Economics* (Northampton, MA: Edward Elgar, 2020); Sarah Myers West, "Data Capitalism: Redefining the Logics of Surveillance and Privacy," *Business & Society* 58, no. 1 (2019): 20–41; Joseph Turow and Nick Couldry, "Media as Data Extraction: Towards a New Map of a Transformed Communications Field," *Journal of Communication* 68, no. 2 (2018): 415–423; Nicole S. Cohen, "The Valorization of Surveillance: Towards a Political Economy of Surveillance," *Democratic Communiqué* 22, no. 1 (2008): 5–22; Christian Fuchs, *Social Media: A Critical Introduction*, 3rd ed. (New York: Routledge, 2021); Kean Birch, "Automated Neoliberalism? The Digital Organisation of Markets in Technoscientific Capitalism," *New Formations* 100–101 (2020): 10–27.

9. Jeremy Wade Morris, "Curation by Code: Infomediaries and the Data Mining of Taste," *European Journal of Cultural Studies* 18, nos. 4–5 (2015): 446–463; Karen Yeung, "'Hypernudge': Big Data as a Mode of Regulation by Design," *Information, Communication & Society* 20, no. 1 (2017): 118–136; Selena Nemorin, *Biosurveillance in New Media Marketing* (Cham, Switzerland: Palgrave Macmillan, 2018); Aaron Darmody and Detlev Zwick, "Manipulate to Empower: Hyper-relevance and the Contradictions of Marketing in the Age of Surveillance Capitalism," *Big Data & Society* 7, no. 1 (2020): 1–12; Andrew McStay, "Micro-moments, Liquidity, Intimacy and Automation: Developments in Programmatic Ad-tech," in *Commercial Communication in the Digital Age—Information or Disinformation?* ed. G. Siegert, M. B. von Rimscha, and S. Grubenmann (Munich: De Gruyter, 2017), 143–159; Emily West, "Amazon: Surveillance as a Service," *Surveillance & Society* 17, nos. 17–18 (2019): 27–33.

10. Wolfie Christl, "How Companies Collect, Combine, Analyze, Trade, and Use Personal Data on Billions," Cracked Labs, June 2017, https://crackedlabs.org/dl/CrackedLabs_Christl_CorporateSurveillance.pdf.

11. Alibaba, Tencent, and ByteDance (TikTok) rank alongside these giants in global advertising revenue.

12. Salomé Viljoen, Jake Goldenfein, and Lee McGuigan, "Design Choices: Mechanism Design and Platform Capitalism," *Big Data & Society* 8, no. 2 (2021): 1–13.

13. John Law, "Technology and Heterogeneous Engineering: The Case of Portuguese Expansion," in *The Social Construction of Technological Systems*, ed. Wiebe E. Bijker, Thomas P. Hughes, and Trevor Pinch (Cambridge, MA: MIT Press, 1987), 111–134.

14. Karen Weise, "Amazon Knows What You Buy: And It's Building a Big Ad Business from It," *New York Times*, January 20, 2019, https://www.nytimes.com/2019/01/20/technology/amazon-ads-advertising.html; Sam Biddle, "Facebook Uses Artificial Intelligence to Predict Your Future Actions for Advertisers, Says Confidential Document," *Intercept*, April 13, 2018, https://theintercept.com/2018/04/13/facebook-advertising-data-artificial-intelligence-ai/; "Google, Mastercard Cut Secret Ad Deal to Track Retail Sales," *Advertising Age*, August 30, 2018, http://adage.com/article/digital/google-mastercard-cut-secret-ad-deal-track-retail-sales/314776.

15. Shoshana Zuboff, *The Age of Surveillance Capitalism: The Fight for a Human Future at the New Frontier of Power* (New York: PublicAffairs, 2019).

16. Ian Leslie, "The Death of Don Draper," *New Statesman*, July 25, 2018, https://www.newstatesman.com/science-tech/internet/2018/07/death-don-draper.

17. Cambridge Analytica, "Don Draper's Dead: Alexander Nix Meets Ogilvy's Rory Sutherland," *CA Commercial* (blog), August 8, 2017, https://ca-commercial.com/news/don-drapers-dead-alexander-nix-meets-ogilvys-rory-sutherland; Ken Auletta, "How the Math Men Overthrew the Mad Men," *New Yorker*, May 21, 2018, https://www.newyorker.com/news/annals-of-communications/how-the-math-men-overthrew-the-mad-men. The gendered term "math men" is problematic, especially when applied to media buying and research, where women have broken barriers to advancement in the male-dominated advertising industry. The term hints, not for the first time, that a discourse of scientism and technical expertise is being used to designate increasingly high-status activities as masculine.

18. "Surveillance capitalism" is usually associated with Zuboff, who claims that Google and other technology companies inaugurated a new logic of capital accumulation. The term is also used by John Bellamy Foster and Robert McChesney to make a more rigorous historical argument. They locate the emergence of surveillance capitalism within a trio of forces that articulated information, computer databases, and communication media: the capitalist sales effort, militarization, and financialization. Zuboff's critique of contemporary conditions has its merits, but Foster and McChesney provide a stronger basis for understanding surveillance capitalism's rise and structure. John Bellamy Foster and Robert W. McChesney, "Surveillance Capitalism: Monopoly-Finance Capital, the Military-Industrial Complex, and the Digital Age," *Monthly Review* 66, no. 3 (2014): 1–31. See also Evgeny Morozov, "Capitalism's New Clothes," *Baffler*, February 4, 2019, https://thebaffler.com/latest/capitalisms-new-clothes-morozov.

19. Dallas W. Smythe and H. H. Wilson, "Cold-War-Mindedness and the Mass Media," in *Struggles against History: United States Foreign Policy in an Age of Revolution*, ed. Neal D. Houghton (New York: Simon & Schuster, 1968), 59–78; Erik Barnouw, *The Sponsor: Notes on a Modern Potentate* (New York: Oxford University Press, 1978); Stuart Ewen, *Captains of Consciousness: Advertising and the Social Roots of the Consumer Culture* (New York: Basic Books, 1976); Raymond Williams, "Advertising: The Magic System," in *Problems in Materialism and Culture* (London: Verso, 1980), 170–195; Susan Strasser, *Satisfaction Guaranteed: The Making of the American Mass Market* (New York: Pantheon

Books, 1989); C. Edwin Baker, *Advertising and a Democratic Press* (Princeton, NJ: Princeton University Press, 1994); George Gerbner, "Cultivation Analysis: An Overview," *Mass Communication & Society* 1, nos. 3–4 (1998): 175–194; Graham Murdock, "Producing Consumerism: Commodities, Ideologies, Practices," in *Critique, Social Media and the Information Society*, ed. Christian Fuchs and Marisol Sandoval (New York: Routledge, 2013), 125–143.

20. Dallas W. Smythe, *Dependency Road: Communications, Capitalism, Consciousness, and Canada* (Norwood, NJ: Ablex, 1981); Susan Smulyan, *Selling Radio: The Commercialization of American Broadcasting 1920–1934* (Washington, DC: Smithsonian Institution Press, 1994); Richard Ohmann, *Selling Culture: Magazines, Markets, and Class at the Turn of the Century* (London: Verso, 1996); Joseph Turow, *Breaking up America: Advertisers and the New Media World* (Chicago: University of Chicago Press, 1997); Mark Andrejevic, *Reality TV: The Work of Being Watched* (Lanham, MD: Rowman & Littlefield, 2004); Eileen R. Meehan, *Why TV Is Not Our Fault* (Lanham, MD: Rowman & Littlefield, 2005); Vincent Manzerolle, "Mobilizing the Audience Commodity: Digital Labour in a Wireless World," *Ephemera* 10, no. 4 (November 2010): 455–469.

For some outstanding works illustrating competing values, see Anna McCarthy, *The Citizen Machine: Governing by Television in 1950s America* (New York: New Press, 2010); David Hesmondhalgh, *The Cultural Industries*, 4th ed. (Los Angeles: Sage, 2019); Angèle Christin, *Metrics at Work: Journalism and the Contested Meaning of Algorithms* (Princeton, NJ: Princeton University Press, 2020).

21. For insightful discussions of distinctive elements, see Detlev Zwick and Janice Denegri Knott, "Manufacturing Customers: The Database as New Means of Production," *Journal of Consumer Culture* 9, no. 2 (2009): 221–247; Adam Arvidsson, "Facebook and Finance: On the Social Logic of the Derivative," *Theory, Culture & Society* 33, no. 6 (2016): 3–23; Adrian Mackenzie, *Machine Learners: Archaeology of a Data Practice* (Cambridge, MA: MIT Press, 2017); Bernhard Rieder, "Scrutinizing an Algorithmic Technique: The Bayes Classifier as an Interested Reading of Reality," *Information, Communication & Society* 20, no. 1 (2017): 100–117; Nick Dyer-Witheford, Atle Mikkola Kjøsen, and James Steinhoff, *Inhuman Power: Artificial Intelligence and the Future of Capitalism* (London: Pluto Press, 2019); Jenna Burrell and Marion Fourcade, "The Society of Algorithms," *Annual Review of Sociology* 47 (2021): 213–247.

22. Inger Stole, "Persistent Pursuit of Personal Information: A Historical Perspective on Digital Advertising Strategies," *Critical Studies in Media Communication* 31, no. 2 (2014): 129–133.

23. On these "cyborg sciences" and the "rationality project" more broadly, see Philip Mirowski, *Machine Dreams: Economics Becomes a Cyborg Science* (New York: Cambridge University Press, 2002); S. M. Amadae, *Rationalizing Capitalist Democracy: The Cold War Origins of Rational Choice Liberalism* (Chicago: University of Chicago Press, 2003).

24. Melvin Anshen, "Management Science in Marketing: Status and Prospects," *Management Science* 2, no. 3 (1956): 224.

25. Wroe Alderson, *Marketing Behavior and Executive Action* (Homewood, IL: Richard D. Irwin, 1957), 408.

26. Martin Kenneth Starr, "Management Science and Marketing Science," *Management Science* 10, no. 3 (1964): 568.

27. For broader but related claims about science and technology, see Dallas W. Smythe, "The Political Character of Science (Including Communication Science), or Science Is Not Ecumenical," in *Communication and Class Struggle*, vol. 1, *Capitalism, Imperialism*, ed. Armand Mattelart and Seth Siegelaub (New York: International General, 1979), 171–176.

28. Ohmann, *Selling Culture*, 13.

29. Smulyan, *Selling Radio*; Robert W. McChesney, *Telecommunications, Mass Media, and Democracy* (New York: Oxford University Press, 1993); Victor Pickard, *America's Battle for Media Democracy: The Triumph of Corporate Libertarianism and the Future of Media Reform* (New York: Cambridge University Press, 2015); Erik Barnouw, *Tube of Plenty*, 2nd rev. ed. (New York: Oxford University Press, 1990); Cynthia B. Meyers, *A Word from Our Sponsor: Admen, Advertising, and the Golden Age of Radio* (New York: Fordham University Press, 2014); Harold A. Innis, "The Press, a Neglected Factor in the Economic History of the Twentieth Century," in *Changing Concepts of Time* (Lanham, MD: Rowman & Littlefield, 2004), 73–104; William Boddy, *Fifties Television: The Industry and Its Critics* (Urbana: University of Illinois Press, 1990); Thomas Streeter, *Selling the Air: A Critique of the Policy of Commercial Broadcasting in the United States* (Chicago: University of Chicago Press, 1996).

30. "TV: Medium Too Good Not to Use," *Broadcasting*, December 9, 1957, 40; emphasis added.

31. Peter Drucker, *The Practice of Management* (New York: Harper, 1954), 371.

32. The phrase "selling the American people" is not meant to conflate the commodification of consumers with the ownership and sale of bodies through slavery or human trafficking. Racialization and long histories of injustice do relate to how marketing systems classify, evaluate, and engage (or not) with individuals or groups. For a superb essay connecting audience commodification to racial capitalism, see Marcel Rosa-Salas, "Making the Mass White: How Racial Segregation Shaped Consumer Segmentation," in *Race in the Marketplace: Crossing Critical Boundaries*, ed. Guillaume D. Johnson, Kevin D. Thomas, Anthony Kwame Harrison, and Sonya A. Grier (Cham, Switzerland: Palgrave Macmillan, 2019), 21–38. On audience commodification, see Dallas W. Smythe, "Communications: Blindspot of Western Marxism," *Canadian Journal of Political and Social Theory* 1, no. 3 (1977): 1–27; Graham Murdock, "Blindspots about Western Marxism: A Reply to Dallas Smythe," *Canadian Journal of Political and Social Theory* 2, no. 2 (1978): 109–119; Eileen R. Meehan, "Ratings and the Institutional Approach: A Third Answer to the Commodity Question," *Critical Studies in Mass Communication* 1, no. 2 (1984): 216–225; Sut Jhally and Bill

Livant, "Watching as Working: The Valorization of Audience Consciousness," *Journal of Communication* 36, no. 3 (1986): 124–143; Ien Ang, *Desperately Seeking the Audience* (New York: Routledge, 1991); Philip M. Napoli, *Audience Economics* (New York: Columbia University Press, 2003); Fernando Bermejo, "Audience Manufacture in Historical Perspective: From Broadcasting to Google," *New Media & Society* 11, nos. 1–2 (2009): 133–154; Zoe Sherman, *Modern Advertising and the Market for Audience Attention* (New York: Routledge, 2020). The terms "attention merchant" and "choice architect" come, respectively, from Tim Wu, *The Attention Merchants: The Epic Scramble to Get Inside Our Heads* (New York: Vintage, 2017), and Richard Thaler and Cass Sunstein, *Nudge: Improving Decisions about Health, Wealth, and Happiness* (New Haven, CT: Yale University Press, 2008).

33. Lizabeth Cohen, *A Consumers' Republic: The Politics of Mass Consumption in Postwar America* (New York: Vintage Books, 2004), 298–305; Sarah E. Igo, *The Averaged American: Surveys, Citizens, and the Making of a Mass Public* (Cambridge, MA: Harvard University Press, 2007); Adam Arvidsson, "On the 'Prehistory of the Panoptic Sort': Mobility in Market Research," *Surveillance & Society* 1, no. 4 (2004): 456–474.

34. Arlene Dávila, *Latinos Inc.: The Marketing and Making of a People*, updated ed. (Berkeley: University of California Press, 2012); Rena Bivens and Oliver L. Haimson, "Baking Gender into Social Media Design: How Platforms Shape Categories for Users and Advertisers," *Social Media + Society* 2, no. 4 (2016): 1–12; Kelley Cotter, Mel Mederios, Chakyung Pak, and Kjerstin Thorson, "'Reach the Right People': The Politics of 'Interests' in Facebook's Classification System for Ad Targeting," *Big Data & Society* 8, no. 1 (2021): 1–16.

35. John B. Watson, "Influencing the Mind of Another" (1935), reprinted from an address delivered to the Montreal Ad Club, J. Walter Thompson Publications Collection, box DG10, John W. Hartman Center for Sales, Advertising & Marketing History, Duke University, Durham, NC (hereafter Hartman Center).

36. Armand Mattelart, *Advertising International: The Privatisation of Public Space*, trans. Michael Chanan (London: Routledge, 1991); Mark Bartholomew, *Adcreep: The Case against Modern Marketing* (Stanford, CA: Stanford University Press, 2017).

37. See Hesmondhalgh, *Cultural Industries*, 112–129.

38. Robert W. McChesney, "The Internet and U.S. Communication Policy-Making in Historical and Critical Perspective," *Journal of Communication* 46, no. 1 (1996): 98–124; Dan Schiller, *Digital Capitalism: Networking the Global Market System* (Cambridge, MA: MIT Press, 1999); Robin Mansell, *Imagining the Internet: Communication, Innovation, and Governance* (New York: Oxford University Press, 2012); Matthew Crain, *Profit over Privacy: How Surveillance Advertising Conquered the Internet* (Minneapolis: University of Minnesota Press, 2021).

39. Turow, *Daily You*; David Golumbia, *The Cultural Logic of Computation* (Cambridge, MA: Harvard University Press, 2009); Robert W. Gehl, *Reverse Engineering Social Media: Software, Culture, and Political Economy in New Media Capitalism* (Philadelphia: Temple

University Press, 2014); Andrew McStay, *The Mood of Information: A Critique of Online Behavioural Advertising* (London: Continuum, 2011); Tanya Kant, *Making It Personal: Algorithmic Personalization, Identity, and Everyday Life* (New York: Oxford University Press, 2020).

40. Zygmunt Bauman and David Lyon, *Liquid Surveillance: A Conversation* (Cambridge: Polity, 2013); Sun-ha Hong, *Technologies of Speculation: The Limits of Knowledge in a Data-Driven Society* (New York: NYU Press, 2019).

41. Crain, *Profit over Privacy*; Srnicek, *Platform Capitalism*; Jathan Sadowski, "When Data Is Capital: Datafication, Accumulation, and Extraction," *Big Data & Society* 6, no. 1 (2019): 1–12.

42. Ryan Calo, "Digital Market Manipulation," *George Washington Law Review* 82, no. 4 (2014): 995–1051; Daniel Susser, Beate Roessler, and Helen Nissenbaum, "Technology, Autonomy, and Manipulation," *Internet Policy Review* 8, no. 2 (2019): 1–22.

43. Safiya Umoja Noble, *Algorithms of Oppression* (New York: NYU Press, 2018), 5.

44. Joshua Braun and Jessica Eklund, "Fake News, Real Money: Ad Tech Platforms, Profit-Driven Hoaxes, and the Business of Journalism," *Digital Journalism* 7, no. 1 (2019): 1–21; Siva Vaidhyanathan, *Antisocial Media: How Facebook Disconnects Us and Undermines Democracy* (New York: Oxford University Press, 2018).

45. Brenton J. Malin, "Advertising as a Tax Expenditure: The Tax Deduction for Advertising and America's Hidden Public Media System," *International Journal of Communication* 8, no. 1 (2020): 2–17.

46. See, e.g., Julia Angwin and Terry Parris Jr., "Facebook Lets Advertisers Exclude Users by Race," *ProPublica*, October 28, 2016, https://www.propublica.org/article /facebook-lets-advertisers-exclude-users-by-race.

47. Muhammad Ali, Piotr Sapiezynski, Miranda Bogen, Aleksandra Korolva, Alan Mislove, and Aaron Rieke, "Discrimination through Optimization: How Facebook's Ad Delivery Can Lead to Biased Outcomes," *Proceedings of the ACM on Human-Computer Interaction* 3 (2019): 1–30.

48. Virginia Eubanks, *Automated Inequality: How High-Tech Tools Profile, Police, and Punish the Poor* (New York: St. Martin's Press, 2017); Cathy O'Neil, *Weapons of Math Destruction* (New York: Crown, 2016).

49. John Cheney-Lippold, *We Are Data: Algorithms and the Making of Our Digital Selves* (New York: NYU Press, 2017); Göran Bolin and Jonas Andersson Schwarz, "Heuristics of the Algorithm: Big Data, User Interpretation, and Institutional Translation," *Big Data & Society* 2, no. 2 (2015): 1–12; Liz Moor and Celia Lury, "Price and the Person: Markets, Discrimination, and Personhood," *Journal of Cultural Economy* 11, no. 6 (2018): 501–513; Eran Fisher and Yoav Mehozay, "How Algorithms See Their Audience: Media Epistemes and the Changing Conception of the Individual," *Media, Culture & Society* 41, no. 8 (2019): 1176–1191.

50. Marcel Rosa-Salas, "Total Market American: Race, Data, and Advertising" (PhD diss., New York University, 2020); Tressie McMillan Cottom, "Where Platform Capitalism and Racial Capitalism Meet: The Sociology of Race and Racism in the Digital Society," *Sociology of Race and Ethnicity* 6, no. 4 (2020): 441–449.

51. Oscar H. Gandy Jr., "Coming to Terms with the Panoptic Sort," in *Computers, Surveillance, and Privacy*, ed. David Lyon and Elia Zureik (Minneapolis: University of Minnesota Press, 1996), 133.

52. José van Dijck, *The Culture of Connectivity: A Critical History of Social Media* (New York: Oxford University Press, 2013); Taina Bucher, *If . . . Then: Algorithmic Power and Politics* (New York: Oxford University Press, 2018); Elinor Carmi, "Rhythmedia: A Study of Facebook Immune System," *Theory, Culture & Society* 37, no. 5 (2020): 119–138.

53. Robert Cluley, "The Politics of Consumer Data," *Marketing Theory* 20, no. 1 (2020): 45–63.

54. Nick Seaver, "Algorithmic Recommendations and Synaptic Functions," *Limn* 2 (2012), https://escholarship.org/uc/item/7g48p7pb; Robert Prey, "Nothing Personal: Algorithmic Individuation on Music Streaming Platforms," *Media, Culture & Society* 40, no. 7 (2018): 1086–1100.

55. See Mark Andrejevic, *iSpy: Surveillance and Power in the Interactive Era* (Lawrence: University Press of Kansas, 2007).

56. Luke Stark, "Algorithmic Psychometrics and the Scalable Subject," *Social Studies of Science* 48, no. 2 (2018): 204–231.

57. See Zuboff, *Age of Surveillance Capitalism*, 87–92.

58. Gandy, "Coming to Terms with the Panoptic Sort," 134.

59. Peter C. Newman, "The Lost Marshall McLuhan Tapes," *Maclean's*, July 16, 2013, https://www.macleans.ca/society/life/the-lost-mcluhan-tapes-2/.

60. Andrea Mennicken and Wendy Nelson Espeland, "What's New with Numbers? Approaches to the Study of Quantification," *Annual Review of Sociology* 45 (2019): 223–245; John Durham Peters, *The Marvelous Clouds: Toward a Philosophy of Elemental Media* (Chicago: University of Chicago Press, 2015); Dan Bouk, "The History and Political Economy of Personal Data over the Last Two Centuries in Three Acts," *Osiris* 32 (2017): 85–106; Theodora Dryer, "Algorithms under the Reign of Probability," *IEEE Annals of the History of Computing* 40, no. 1 (2018): 93–96; Josh Lauer and Kenneth Lipartito, eds., *Surveillance Capitalism in America* (Philadelphia: University of Pennsylvania Press, 2021); Justin Joque, *Revolutionary Mathematics: Artificial Intelligence, Statistics and the Logic of Capitalism* (New York: Verso, 2022).

61. Simone Browne, *Dark Matters: On the Surveillance of Blackness* (Durham, NC: Duke University Press, 2015); Caitlin Rosenthal, *Accounting for Slavery: Masters and Management* (Cambridge, MA: Harvard University Press, 2018).

62. Lauer, *Creditworthy*; Dan Bouk, *How Our Days Became Numbered: Risk and the Rise of the Statistical Individual* (Chicago: University of Chicago Press, 2015).

63. Caitlin Zaloom, *Out of the Pits: Traders and Technology from Chicago to London* (Chicago: University of Chicago Press, 2006).

64. Thomas P. Hughes, *Human-Built World: How to Think about Technology and Culture* (Chicago: University of Chicago Press, 2004), 77–110.

65. Vincent Mosco, "Introduction: Information in the Pay-Per Society," in *The Political Economy of Information*, ed. Vincent Mosco and Janet Wasko (Madison: University of Wisconsin Press, 1988), 4.

66. Tim Hwang, *Subprime Attention Crisis: The Timebomb at the Heart of the Internet* (New York: FSG, 2020).

67. Comment from an audience member, "Advanced TV Ads: The Rise to Scale," Advanced Advertising Summit, New York City, March 27, 2017.

68. Andrew McStay, *Emotional AI: The Rise of Empathic Media* (Los Angeles: Sage, 2018); Aaron Shapiro, *Design, Control, Predict: Logistical Governance in the Smart City* (Minneapolis: University of Minnesota Press, 2020).

69. Mattelart, *Advertising International*, 170.

CHAPTER 1

1. Ethan Cramer-Flood, "Worldwide Digital Ad Spending Year-End Update," *eMarketer*, November 23, 2021, https://www.emarketer.com/content/worldwide-digital-ad -spending-year-end-update; Sara Lebow, "Google, Facebook, and Amazon to Account for 64% of US Digital Ad Spending This Year," *eMarketer*, November 3, 2021, https:// www.emarketer.com/content/google-facebook-amazon-account-over-70-of-us-digital -ad-spending.

2. Tim Hwang, *Subprime Attention Crisis: The Timebomb at the Heart of the Internet* (New York: FSG, 2020), 116.

3. Ronan Shields, "Apple's Latest Privacy Announcement Could Be More Impactful than CCPA or GDPR," *Adweek*, June 23, 2020; Natasha Lomas, "On Meta's 'Regulatory Headwinds' and Adtech's Privacy Reckoning," *TechCrunch*, February 4, 2022, https://techcrunch.com/2022/02/04/on-metas-regulatory-headwinds-and-adtechs -privacy-reckoning/.

4. Further details about how adtech works are available in a growing body of cross-disciplinary research. See, e.g., Joseph Turow, *The Daily You: How the New Advertising Industry Is Defining Your Identity and Your Worth* (New Haven, CT: Yale University Press, 2011); Robert W. Gehl, *Reverse Engineering Social Media: Software, Culture, and Political Economy in New Media Capitalism* (Philadelphia: Temple University Press, 2014); Andrew McStay, "Digital Advertising and Adtech: Programmatic Platforms, Identity and Moments," in *The Advertising Handbook*, 4th ed., ed. Jonathan Hardy, Helen Powell, and Iain MacRury (New York: Routledge, 2018), 88–101; Ramon Lobato and Julian Thomas, "Formats and Formalization in Internet Advertising," in *Format Matters*, ed. Marek Jancovic, Axel Volmar, and Alexandra Schneider (Lüneburg, Germany:

Meson Press, 2020), 65–80; Christina Alaimo and Jannis Kallinikos, "Objects, Metrics and Practices: An Inquiry into the Programmatic Advertising Ecosystem," in *Living with Monsters? Social Implications of Algorithmic Phenomena, Hybrid Agency, and the Performativity of Technology*, ed. Ulrike Schultze, Margunn Aanestad, Magnus Mähring, Carsten Østerlund, and Kai Riemer (Cham, Switzerland: Springer, 2018), 110–123; Muhammad Ahmad Bashir, "On the Privacy Implications of Real Time Bidding" (PhD diss., Northeastern University, 2019); Reuben Binns, "Tracking on the Web, Mobile, and the Internet-of-Things," accessed April 29, 2022, https://arxiv.org/pdf/2201.10831.pdf.

5. Gabriel Nicholas and Aaron Shapiro, "Failed Hybrids: The Death and Life of Bluetooth Proximity Marketing," *Mobile Media & Communication* 9, no. 3 (2020): 465–487.

6. Jack Marshall, "WTF is Programmatic Advertising?," *Digiday*, February 20, 2014, https://digiday.com/media/what-is-programmatic-advertising/.

7. William Ammerman, *The Invisible Brand: Marketing in the Age of Automation, Big Data, and Machine Learning* (New York: McGraw-Hill, 2019), 30.

8. "About Value Optimization," Meta Business Help Center, accessed April 29, 2022, https://www.facebook.com/business/help/296463804090290?id=561906377587030.

9. Aaron Shapiro, *Design, Control, Predict: Logistical Governance in the Smart City* (Minneapolis: University of Minnesota Press, 2020), 64–71.

10. Chris Kane, "How to Train Your DSP," *Programmatic I/O*, 2018, https://jouncemedia .com/blog/2018/4/12/catch-the-replay-how-to-train-your-dsp.

11. Fernando N. van der Vlist and Anne Helmond, "How Partners Mediate Platform Power: Mapping Business and Data Partnerships in the Social Media Ecosystem," *Big Data & Society* 8, no. 1 (2021): 1–16.

12. Alexandra Bruell, "Inside the Hidden Costs of Programmatic," *Advertising Age* 86, no. 17 (2015): 14.

13. Nick Kostov and David-Gauthier-Villars, "Digital Revolution Upends Ad Industry—A Divide between Old Guard and New Tech Hires," *Wall Street Journal*, January 20, 2018, A1.

14. John Sinclair, "Advertising and Media in the Age of the Algorithm," *International Journal of Communication* 10 (2016): 3522–3535.

15. John Deighton, "Rethinking the Profession Formerly Known as Advertising," *Journal of Advertising Research* 58, no. 4 (2017): 357–361.

16. Dina Srinivasan, "Why Google Dominates Advertising Markets," *Stanford Technology Law Review* 24, no. 1 (2020): 55–175.

17. Srinivasan, 101.

18. Srinivasan, 122–127.

19. Gilad Edelman, "Google's Alleged Scheme to Corner the Online Ad Market," *Wired*, January 14, 2022, https://www.wired.com/story/google-antitrust-ad-market-lawsuit/.

20. Claire Melford and Craig Fagan, "Cutting the Funding of Disinformation: The Ad-Tech Solution," Global Disinformation Index, May 2019, https://disinformationindex .org/wp-content/uploads/2019/05/GDI_Report_Screen_AW2.pdf.

21. Ryan Barwick, "Brands Are Still Playing Ball with Clickbait Ad Sites, Advertising's Roach that Will Survive the Bomb," *Morning Brew*, September 8, 2021, https://www .morningbrew.com/marketing/stories/2021/09/08/brands-still-playing-ball-clickbait -ad-sites-advertisings-roach-will-survive-bomb.

22. Caitlin Petre, "The Traffic Factories: Metrics at Chartbeat, Gawker Media, and the New York Times," *Tow Center for Digital Journalism*, May 7, 2015, https:// academiccommons.columbia.edu/doi/10.7916/D80293W1. Relatedly, on "content factories," see Nicole S. Cohen, *Writers' Rights: Freelance Journalism in a Digital Age* (Montreal, QC: McGill-Queens University Press, 2016), chap. 5.

23. Oscar H. Gandy Jr., *The Panoptic Sort: A Political Economy of Personal Information*, 2nd ed. (New York: Oxford University Press, 2021).

24. Donald MacKenzie, "Cookies, Pixels and Fingerprints," *London Review of Books* 43, no. 7 (2021): 31–34.

25. "Identity Graphs on AWS," accessed October 13, 2021, https://aws.amazon.com /neptune/identity-graphs-on-aws/.

26. Tarleton Gillespie, "The Relevance of Algorithms" in *Media Technologies: Essays on Communication, Materiality, and Society*, ed. Tarleton Gillespie, Pablo J. Boczkowski, and Kirsten A. Foot (Cambridge, MA: MIT Press, 2016), 167–194; Angèle Christin, *Metrics at Work: Journalism and the Contested Meaning of Algorithms* (Princeton, NJ: Princeton University Press, 2020).

27. Kane, "How to Train Your DSP."

28. Solon Barocas and Helen Nissenbaum, "Big Data's End Run around Anonymity and Consent," in *Privacy, Big Data, and the Public Good*, ed. Julia Lane, Victoria Stodden, Stefan Bender, and Helen Nissenbaum (New York: Cambridge University Press, 2014), 44–75; Michael Veale and Frederik Zuiderveen Borgesius, "Adtech and Real-Time Bidding under European Data Protection Law," *German Law Journal* 23, no. 2 (2022): 226–256.

29. "Dr. Johnny Ryan Testimony," *Understanding the Digital Advertising Ecosystem and the Impact of Data Privacy and Competition Policy*, US Senate Committee on the Judiciary, May 21, 2019, Washington, DC, https://www.judiciary.senate.gov/imo/media /doc/Ryan%20Testimony.pdf.

30. Joseph Cox, "How the U.S. Military Buys Location Data from Ordinary Apps," *Motherboard*, November 16, 2020, https://www.vice.com/en/article/jgqm5x/us-military -location-data-xmode-locate-x; Byron Tau and Georgia Wells, "Grindr User Data Was Sold through Ad Networks," *Wall Street Journal*, May 2, 2022, https://www.wsj .com/articles/grindr-user-data-has-been-for-sale-for-years-11651492800.

31. Joseph Cox, "Data Broker Is Selling Location Data of People Who Visit Abortion Clinics," *Vice*, May 3, 2022, https://www.vice.com/en/article/m7vzjb/location-data -abortion-clinics-safegraph-planned-parenthood.

32. Gilad Edelman, "Follow the Money: How Digital Ads Subsidize the Worst of the Web," *Wired*, July 28, 2020, https://www.wired.com/story/how-digital-ads-subsidize -worst-web/.

33. Shoshana Wodinsky, "Target Is Sponsoring News Stories of Its Own Destruction— and Likely Has No Idea," *Gizmodo*, May 26, 2020, https://gizmodo.com/target-is -sponsoring-news-stories-of-its-own-destructio-1843759747.

34. Sophie Bishop, "Influencer Management Tools: Algorithmic Cultures, Brand Safety, and Bias," *Social Media + Society* 7, no. 1 (2021): 1–13; Robyn Caplan and Tarleton Gillespie, "Tiered Governance and Demonetization: The Shifting Terms of Labor and Compensation in the Platform Economy," *Social Media + Society* 6, no. 2 (2020): 1–13.

35. Shoshana Wodinksy, "The Newest Chrome Update Is about Power, Not Privacy," *Gizmodo*, May 20, 2020, https://gizmodo.com/the-newest-chrome-update-is-about -power-not-privacy-1843546136.

36. Jérôme Bourdon and Cécile Méadel, "Globalizing the Peoplemetered Audience," in *The Routledge Companion to Global Television*, ed. Shawn Shimpach (New York: Routledge, 2019), 121–129.

37. This section builds on a large body of work, including Todd Gitlin, *Inside Prime Time* (New York: Pantheon Books, 1983); Eileen R. Meehan, "Ratings and the Institutional Approach: A Third Answer to the Commodity Question," *Critical Studies in Mass Communication* 1, no. 2 (1984): 216–225; Ien Ang, *Desperately Seeking the Audience* (New York: Routledge, 1991); Philip Napoli, *Audience Economics* (New York: Columbia University Press, 2003); Jérôme Bourdon and Cécile Méadel, "Inside Television Audience Measurement: Deconstructing the Ratings Machine," *Media, Culture & Society* 33, no. 5 (2011): 791–800.

38. Dallas W. Smythe, *Dependency Road: Communications, Capitalism, Consciousness, and Canada* (Norwood, NJ: Ablex, 1981), chap. 2.

39. Harold A. Innis, *The Bias of Communication*, 2nd ed. (Toronto: University of Toronto Press, 2008), xliii.

40. Fernando Bermejo, *The Internet Audience: Constitution and Measurement* (New York: Peter Lang, 2007).

41. On continuities in audience measurement and the work of self-surveillance, see Jennifer Hessler, "Peoplemeter Technologies and the Biometric Turn in Audience Measurement," *Television & New Media* 22, no. 4 (2021): 400–419.

42. Matthew Crain, *Profit over Privacy: How Surveillance Advertising Conquered the Internet* (Minneapolis: University of Minnesota Press, 2020), 67.

43. Companies like Meta also produce evidence of behavior for an audience of investors who carefully monitor user metrics.

44. Josh Lauer, "Surveillance History and the History of New Media: An Evidential Paradigm," *New Media & Society* 14, no. 4 (2011): 568.

45. Lauer, 579.

46. See Michael Buckland, "What Is a 'Document'?," *Journal of the American Society for Information Science* 48, no. 9 (1997): 806.

47. I am drawing on Bronwyn Parry, *Trading the Genome: Investigating the Commodification of Bio-Information* (New York: Columbia University, 2004).

48. Astrid Mager, "Algorithmic Ideology: How Capitalist Society Shapes Search Engines," *Information, Communication & Society* 15, no. 5 (2012): 769–787; Micky Lee, "Google Ads and the Blindspot Debate," *Media, Culture & Society* 33, no. 3 (2011): 443–447; Bernhard Rieder and Guillame Sire, "Conflicts of Interest and Incentives to Bias: A Microeconomic Critique of Google's Tangled Position on the Web," *New Media & Society* 16, no. 2 (2013): 195–221; Carolin Gerlitz and Anne Helmond, "The Like Economy: Social Buttons and the Data-Intensive Web," *New Media & Society* 15, no. 8 (2013): 1348–1365; Ramon Lobato, "The Cultural Logic of Digital Intermediaries: YouTube Multichannel Networks," *Convergence* 22, no. 4 (2016): 348–360; David B. Nieborg and Thomas Poell, "The Platformization of Cultural Production: Theorizing the Contingent Cultural Commodity," *New Media & Society* 20, no. 11 (2018): 4275–4292; Elizabeth Van Couvering, "Faces and Charts: Platform Strategies for Visualising the Audience, the Case of Facebook," *Information, Communication & Society* 25, no. 11 (2022): 1524–1541.

49. On the institutionally effective audience, see James S. Ettema and D. Charles Whitney, "The Money Arrow: An Introduction to Audiencemaking," in *Audiencemaking: How the Media Create the Audience*, ed. James S. Ettema and D. Charles Whitney (Thousand Oaks, CA: Sage, 1994), 1–18. For discussions of framing and markets, see Michel Callon, "Introduction: The Embeddedness of Economic Markets in Economics," *Sociological Review* 46, no. S1 (1998): 1–57; Michel Callon and Fabian Muniesa, "Peripheral Vision: Economic Markets as Calculative Collection Devices," *Organization Studies* 26, no. 8 (2005): 1229–1250; N. Anand and Richard A. Peterson, "When Market Information Constitutes Fields: Sensemaking of Markets in the Commercial Music Industry," *Organization Science* 11, no. 3 (2000): 270–284.

50. Vincent Manzerolle and Michael Daubs, "Friction-Free Authenticity: Mobile Social Networks and Transactional Affordances," *Media, Culture & Society* 43, no. 7 (2021): 1279–1296.

51. Garett Sloane, "Instagram Gives Brands New Way to Sell in 'Collection' Ads," *Advertising Age*, February 6, 2018, https://adage.com/article/digital/instagram-brands-sell-collection-ads/312273; Matthew Schneier, "Instagram Introduces Shoppable Influencers," *New York Times*, April 30, 2019, https://www.nytimes.com/2019/04/30/style/instagram-introduces-shoppable-influencers.html; Alison Weissbrot, "Google Extends Shopping Ads to YouTube," *AdExchanger*, November 5, 2019, https://www.adexchanger.com/platforms/google-extends-shopping-ads-to-youtube/; Scott Nover, "TikTok Makes Moves in Social Commerce with Shopify Integration," *Adweek*,

October 27, 2020, https://www.adweek.com/programmatic/tiktok-makes-moves-in
-social-commerce-with-shopify-integration/.

52. Mark Andrejevic, *Automated Media* (New York: Routledge, 2020), 97.

53. Garett Sloane, "How TikTok, Instagram and Snapchat Sell Everything from Ranch
Dressing to Lip Gloss," *Advertising Age*, May 18, 2021, https://adage.com/article/media
/how-tiktok-instagram-and-snapchat-sell-everything-ranch-dressing-lip-gloss/2335766.

54. Nicole Perrin, "Digital Identity Crisis Will Be a Boon for Retail Media Sellers,"
eMarketer, October 4, 2020, https://www.emarketer.com/content/digital-identity
-crisis-will-boon-retail-media-sellers.

55. Stephen Fox, *The Mirror Makers: A History of American Advertisers and Its Creators*
(Urbana: University of Illinois Press, 1994).

56. Adam Arvidsson, "On the 'Prehistory of the Panoptic Sort': Mobility in Market
Research," *Surveillance & Society* 1, no. 4 (2004): 456–474; Jason Pridmore and Detlev
Zwick, "Marketing and the Rise of Commercial Consumer Surveillance," *Surveillance &
Society* 8, no. 3 (2011): 269–277.

CHAPTER 2

Epigraph: Wroe Alderson and Stanley J. Shapiro, eds., *Marketing and the Computer*
(Englewood Cliffs, NJ: Prentice-Hall, 1963), 2.

1. Thomas Frank, *The Conquest of Cool: Business Culture, Counterculture, and the Rise of
Hip Consumerism* (Chicago: University of Chicago Press, 1997).

2. Joseph Turow, *Breaking up America: Advertisers and the New Media World* (Chicago:
University of Chicago Press, 1997); Lizabeth Cohen, *A Consumers' Republic: The Politics of Mass Consumption in Postwar America* (New York: Vintage Books, 2004).

3. On the status of data science in marketing today, see Ken Auletta, *Frenemies: The
Epic Disruption of the Ad Business (and Everything Else)* (New York: Penguin Press,
2018); Steve Lohr, *Data-ism: The Revolution Transforming Decision Making, Consumer
Behavior, and Almost Everything Else* (New York: HarperCollins, 2015).

4. See, e.g., Shoshana Zuboff, *The Age of Surveillance Capitalism* (New York: Public-
Affairs, 2019). My analysis supports the argument in John Bellamy Foster and Robert
W. McChesney, "Surveillance Capitalism: Monopoly-Finance Capital, the Military-
Industrial Complex, and the Digital Age," *Monthly Review* 66, no. 3 (2014): 1–31. See
also Kevin Robins and Frank Webster, "Cybernetic Capitalism: Information, Technology, Everyday Life," in *The Political Economy of Information*, ed. Vincent Mosco
and Janet Wasko (Madison: University of Wisconsin Press, 1988), 44–75.

5. Pamela Walker Laird, *Advertising Progress: American Business and the Rise of Consumer
Marketing* (Baltimore: Johns Hopkins University Press, 1998); Roland Marchand,
Advertising the American Dream: Making Way for Modernity, 1920–1940 (Berkeley: Uni-
versity of California Press, 1985); Susan Strasser, *Satisfaction Guaranteed: The Making*

of the American Mass Market (New York: Pantheon Books, 1989); Jackson Lears, *Fables of Abundance: A Cultural History of Advertising in America* (New York: Basic Books, 1994); Stephen Leiss, William Klein, Sut Jhally, Jacqueline Botterill, and Kyle Asquith, *Social Communication in Advertising*, 4th ed. (New York: Routledge, 2018); Stephen Fox, *The Mirror Makers: A History of American Advertisers and Its Creators* (Urbana: University of Illinois Press, 1994); Daniel Pope, *The Making of Modern Advertising* (New York: Basic Books, 1983); Peggy J. Kreshel, "The 'Culture' of J. Walter Thompson, 1915–1925," *Public Relations Review* 16, no. 3 (1990): 80–93; Peggy J. Kreshel, "John B. Watson at J. Walter Thompson: The Legitimation of 'Science' in Advertising," *Journal of Advertising* 19, no. 2 (1990): 49–59.

6. Joseph Turow, *The Daily You: How the New Advertising Industry Is Defining Your Identity and Your Worth* (New Haven, CT: Yale University Press, 2011), 19.

7. Michael Schudson, *Advertising, the Uneasy Persuasion* (New York: Basic Books, 1984), chap. 2.

8. Gary L. Lilien and Philip Kotler, *Marketing Decision Making: A Model-Building Approach* (New York: Harper & Row, 1983), 511. Notably, that researcher was working for the Scott Paper Company, which might have been happy to see agencies sink more money into the production of ads.

9. Geoffrey Bowker and Susan Leigh Star, *Sorting Things Out: Classification and Its Consequences* (Cambridge, MA: MIT Press, 1999), 53.

10. On how precarious relationships can increase the status of quantification, see Bruce G. Carruthers and Wendy Nelson Espeland, "Accounting for Rationality: Double-Entry Bookkeeping and the Rhetoric of Economic Rationality," *American Journal of Sociology* 97, no. 1 (1991): 31–69; Theodore Porter, *Trust in Numbers: The Pursuit of Objectivity in Science and Public Life* (Princeton, NJ: Princeton University Press, 1995).

11. Caitlin Rosenthal, *Accounting for Slavery: Masters and Management* (Cambridge, MA: Harvard University Press, 2018).

12. William Leach, *Land of Desire: Merchants, Power, and the Rise of a New American Culture* (New York: Vintage Books, 1994), 35–38.

13. James R. Beniger, *The Control Revolution: Technological and Economic Origins of the Information Society* (Cambridge, MA: Harvard University Press, 1986); Alfred D. Chandler, *The Visible Hand: The Managerial Revolution in American Business* (Cambridge, MA: Belknap Press of Harvard University Press, 1977).

14. Harold A. Innis, *Political Economy in the Modern State*, ed. Robert E. Babe and Edward A. Comor (1946; reprint, Toronto: University of Toronto, 2018), 89.

15. Innis, 89.

16. Dan Schiller, *Digital Depression: Information Technology and Economic Crisis* (Urbana: University of Illinois Press, 2014), 18.

17. Laird, *Advertising Progress*, 254–260.

18. See the works cited above in note 5.

19. Roger Barton, *Advertising Agency Operations and Management* (New York: McGraw-Hill, 1955), 3–4.

20. Zoe Sherman, *Modern Advertising and the Market for Audience Attention* (New York: Routledge, 2020), 36.

21. *Printers' Ink* 14, no. 2 (1896): 80.

22. Fox, *Mirror Makers*, 29–31.

23. Quoted in Laird, *Advertising Progress*, 284.

24. Richard Ohmann, *Selling Culture: Magazines, Markets, and Class at the Turn of the Century* (London: Verso, 1996), 112–113.

25. Innis, *Political Economy in the Modern State*, 27.

26. Laird, *Advertising Progress*, 254.

27. Douglas B. Ward, "Capitalism, Early Market Research, and the Creation of the American Consumer," *Journal of Historical Research in Marketing* 1, no. 2 (2009): 200–223.

28. Richard K. Popp, "The Information Bazaar: Mail-Order Magazines and the Gilded Age Trade in Consumer Data," in *Surveillance Capitalism in America*, ed. Josh Lauer and Kenneth Lipartito (Philadelphia: University of Pennsylvania Press, 2021), 46–64.

29. Daniel Robinson, *The Measure of Democracy* (Toronto: University of Toronto Press, 1999); Sarah E. Igo, *The Averaged American: Surveys, Citizens, and the Making of a Mass Public* (Cambridge, MA: Harvard University Press, 2008); Marcel Rosa-Salas, "Making the Mass White: How Racial Segregation Shaped Consumer Segmentation," in *Race in the Marketplace: Crossing Critical Boundaries*, ed. Guillaume D. Johnson, Kevin D. Thomas, Anthony Kwame Harrison, and Sonya A. Grier (Cham, Switzerland: Palgrave Macmillan, 2019), 21–38.

30. Ian Hacking, "Making up People," in *Reconstructing Individualism*, ed. Thomas C. Heller, Morton Sosna, and David Wellbery (Stanford, CA: Stanford University Press, 1986), 222–236.

31. Stefan Schwarzkopf, "In Search of the Consumer: The History of Market Research from 1890 to 1960," in *The Routledge Companion to Marketing History*, ed. D. G. Brian Jones and Mark Tadajewski (New York: Routledge, 2016), 61–83.

32. Earnest Elmo Calkins and Ralph Holden, *Modern Advertising* (New York: D. Appleton, 1907), 261. Quoted also in Strasser, *Satisfaction Guaranteed*, 147–148.

33. James B. Griffith, *Advertising and Sales Organization* (Chicago: American School of Correspondence, 1909), 3.

34. JoAnne Yates, *Control through Communication: The Rise of System in American Management* (Baltimore: Johns Hopkins University Press, 1989).

35. Sherman, *Modern Advertising and the Market for Audience Attention*, 121–125.

36. Popp, "The Information Bazaar," 47.

37. Richard K. Popp, "Making Advertising Material: Checking Departments, Systematic Reading, and Geographic Order in Nineteenth-Century Advertising," *Book History* 14 (2011): 58–87.

38. Griffith, *Advertising and Sales Organization*, 23.

39. Hebert Newton Casson, *Ads and Sales: A Study of Advertising and Selling, from the Standpoint of the New Principles of Scientific Management* (Chicago: A. C. McClung, 1911), 6–7.

40. Casson, 9–11.

41. Merle Curti, "The Changing Concept of 'Human Nature' in the Literature of American Advertising," *Business History Review* 41, no. 4 (1967): 342.

42. Curti, 347.

43. Laird, *Advertising Progress*, 279.

44. Pope, *Making Modern Advertising*, 141; Beniger, *Control Revolution*, 379, 384.

45. Fox, *Mirror Markers*, 60–61.

46. Fox, 61.

47. Calkins and Holden, *Modern Advertising*, 262–270.

48. Schwarzkopf, "In Search of the Consumer," 66.

49. Daniel J. Robinson, "Mail-Order Doctors and Market Research, 1890–1930," in *The Rise of Marketing and Market Research*, ed. Harmut Berghoff, Philip Scranton, and Uwe Spiekermann (New York: Palgrave Macmillan, 2012), 73–93.

50. Sherman, *Modern Advertising and the Market for Audience Attention*, 110–111.

51. Schwarzkopf, "In Search of the Consumer"; Pope, *Making Modern Advertising*, 141–143.

52. Martin Kenneth Starr, "Management Science and Marketing Science," *Management Science* 10, no. 3 (1964): 557–558.

53. Starr, 558.

54. Wroe Alderson, "Charles Coolidge Parlin," *Journal of Marketing* 21, no. 1 (1956): 1.

55. Alderson, 1.

56. Ward, "Capitalism, Early Market Research, and the Creation of the American Consumer," 209–210.

57. Beniger, *Control Revolution*, 387.

58. Marchand, *Advertising the American Dream*, chaps. 2, 3.

59. E. G. Platt, "Data Publishers Should Furnish Agents," *Printers' Ink*, January 29, 1914, 50.

60. Kreshel, "'Culture' of J. Walter Thompson," 83.

61. Kreshel, "John B. Watson at J. Walter Thompson," 53.

62. Marchand, *Advertising the American Dream*, 26.

63. Fox, *Mirror Makers*, 85.

64. Paul T. Cherington, "Statistics in Market Studies," *Annals of the American Academy of Political and Social Science* 115 (1924): 132.

65. Jan Logemann, "Consumer Engineering and the Rise of Marketing Knowledge, 1920s–1970s," *History of Knowledge*, June 27, 2019, https://historyofknowledge.net/2019/06/27/consumer-engineering-and-the-rise-of-marketing-knowledge-1920s-1970s/

66. See Franck Cochoy, "Another Discipline for the Market Economy: Marketing as a Performative Knowledge and Know-how for Capitalism," *Sociological Review* 46, no. S1 (1998): 205–209.

67. "What the J. Walter Thompson Company Stands For," June 1955, 1, JWT Information Center Records, box 8, Hartman Center.

68. Kreshel, "'Culture' of J. Walter Thompson," 85.

69. Henry C. Link, *The New Psychology of Selling and Advertising* (New York: Macmillan, 1932), 248.

70. Link, 267.

71. *A Report of the Eleventh Meeting of the ARF Operations Research Discussion Group* (New York: Advertising Research Foundation, 1964), 30.

72. Leo Bogart, "Is It Time to Discard the Audience Concept?," *Journal of Marketing* 30, no. 1 (1966): 49.

73. Igo, *Averaged American*, 113.

74. Todd Gitlin, "Media Sociology: The Dominant Approach," *Media, Culture & Society* 6, no. 2 (1978): 233–240.

75. Darrell Lucas quoted in "Mathematical Programming for Better Selection of Advertising Media," *Computers and Automation* 10, no. 12 (1961): 21.

76. Wadsworth H. Mullen, "A Money Measure of Magazine Reader Interest," *Harvard Business Review* 14, no. 3 (1936): 371.

77. Darrell Blaine Lucas and Steuart Henderson Britt, *Measuring Advertising Effectiveness* (New York: McGraw-Hill, 1963), 270.

78. Lucas and Britt, 270–272.

79. Lucas and Britt, 281.

80. Charles Sinclair, "Arrival of Arbitron System Puts Madison Ave. into Tizzy," *Billboard*, December 23, 1957, 2.

81. "Should Agencies Get 20% for Spot TV?," *Broadcasting*, April 28, 1969, 20.

82. Karen Buzzard, *Tracking the Audience: The Ratings Industry from Analog to Digital* (New York: Routledge, 2012); Philip M. Napoli, *Audience Economics* (New York: Columbia University Press, 2003); Ien Ang, *Desperately Seeking the Audience* (New York: Routledge, 1991); Eileen R. Meehan, *Why TV Is Not Our Fault* (Lanham, MD: Rowman & Littlefield, 2005).

83. Buzzard, *Tracking the Audience*, 22.

84. Vance Packard, *The Hidden Persuaders* (New York: Pocket Books, 1957).

85. Ronald A. Fullerton, "The Birth of Consumer Behavior: Motivation Research in the 1940s and 1950s," *Journal of Historical Research in Marketing* 5, no. 2 (2013): 212–222.

86. Robert Vincent Zacher, *Advertising Techniques and Management* (Homewood, IL: R. D. Irwin, 1961), 89.

87. Michael Halbert, "Empirical Research in Consumer Shopping and Motivation," in *Proceedings of the Thirty-Seventh National Conference of the American Marketing Association, New York, December 27th–29th 1955* (New York: American Marketing Association, 1955), 15.

88. Beniger, *Control Revolution*.

CHAPTER 3

Epigraphs: Marion Harper Jr., "A New Profession to Aid Management," *Journal of Marketing* 25, no. 3 (1961): 1; William J. Horvath, "Operations Research—A Scientific Basis for Executive Decisions," *American Statistician* 2, no. 5 (1948): 18.

1. Marshall McLuhan, *The Mechanical Bride: Folklore of Industrial Man* (New York: Vanguard Press, 1951), v.

2. "Ivied Towers Spawn New Breed of Automation Men Searching Ad Answers," *Advertising Age*, December 1962, 3, 36.

3. Jill Lepore, *If Then: How the Simulmatics Corporation Invented the Future* (New York: Liveright, 2020); Fenwick McKelvey, "The Other Cambridge Analytics: Early 'Artificial Intelligence' in American Political Science," in *The Cultural Life of Machine Learning: An Incursion into Critical AI Studies*, ed. Jonathan Roberge and Michael Castelle (London: Palgrave Macmillan, 2020), 117–142.

4. Thomas B. Morgan, "The People-Machine," *Harper's Magazine*, January 1, 1961, 53–57.

5. The widely used term "math men" is problematic, but it is more appropriate in this context than it is today. Although women climbed the ranks in agency research and media departments, they were *not* included among the mathematical experts coming from operations research and management science. In the 1950s and 1960s

the Advertising Research Foundation convened an operations research discussion group, with more than forty members at its peak. As far as I can tell from the records of their meetings, no women participated—except perhaps as the stenographers who produced those records.

6. "Ivied Towers," 3.

7. "Y&R, BBDO Unleash Media Computerization," *Advertising Age*, October 1, 1962, 118.

8. Tim Wu, *The Attention Merchants: The Epic Scramble to Get Inside Our Heads* (New York: Vintage, 2017), 105.

9. James W. Cortada, *The Digital Hand*, vol. 2, *How Computers Changed the Work of American Financial, Telecommunications, Media, and Entertainment Industries* (New York: Oxford University Press, 2004), 353.

10. "Nielsen Reports," *Broadcasting*, September 6, 1948, 29, 34. Eckert and Mauchly's fledgling company failed to fill Nielsen's initial order due to internal problems. Nancy Stern, "The Eckert-Mauchly Computers: Conceptual Triumphs, Commercial Tribulations," *Technology & Culture* 23, no. 4 (October 1982): 579–582.

11. For more on ENIAC, see Jennifer Light, "When Computers Were Women," *Technology & Culture* 40, no. 3 (1999): 455–483; Paul E. Ceruzzi, *Computing: A Concise History* (Cambridge, MA: MIT Press, 2012), 46–48.

12. "Intra-Computer Testing by 2012 Seen by Inventor," *Advertising Age*, December 10, 1962, 89.

13. "As [Dallas] Smythe put it, 'the shotgun is being superseded by rifles.'" William H. Melody, "Audiences, Commodities, and Market Relations: An Introduction to the Audience Commodity Thesis," in *The Audience Commodity in a Digital Age*, ed. Lee McGuigan and Vincent Manzerolle (New York: Peter Lang, 2014), 27.

14. Peter Langhoff, "The Setting: Some Non-Metric Observations," in *Models, Measurement, and Marketing*, ed. Peter Langhoff (Englewood Cliffs, NJ: Prentice-Hall, 1965), 5.

15. Ronald R. Kline, "Cybernetics, Management Science, and Technology Policy: The Emergence of 'Information Technology' as a Keyword, 1948–1985," *Technology & Culture* 47, no. 3 (2006): 521.

16. Nathan Ensmenger, *The Computer Boys Take Over: Computers, Programmers, and the Politics of Technical Expertise* (Cambridge, MA: MIT Press, 2010); Thomas Haigh, "Inventing Information Systems: The Systems Men and the Computer, 1950–1968," *Business History Review* 75, no. 1 (2001): 15–61.

17. Daniel Bell, "The Social Framework of the Information Society," in *The Computer Age: A Twenty-Year View*, ed. Michael L. Dertouzos and Joel Moses (Cambridge, MA: MIT Press, 1979), 166–167.

18. "The RCA 301 Computer," *J. Walter Thompson Company News* 18, no. 11 (April 5, 1963), JWT Staff Newsletters, box MN12, Hartman Center.

19. Caitlin Zaloom, *Out of the Pits: Traders and Technology from Chicago to London* (Chicago: University of Chicago Press, 2006), 165.

20. Philip Mirowski, *Machine Dreams: Economics Becomes a Cyborg Science* (New York: Cambridge University Press, 2002), 177.

21. For important exceptions, see Franck Cochoy, "Another Discipline for the Market Economy: Marketing as a Performative Knowledge and Know-How for Capitalism," *Sociological Review* 46, no. 1 supplement (1998): 194–221; Adam Arvidsson, "Facebook and Finance: On the Social Logic of the Derivative," *Theory, Culture & Society* 33, no. 6 (2016): 3–23; Bernhard Rieder, "Scrutinizing an Algorithmic Technique: The Bayes Classifier as an Interested Reading of Reality," *Information, Communication & Society* 20, no. 1 (2017): 100–117; Adrian Mackenzie, "Personalization and Probabilities: Impersonal Propensities in Online Grocery Shopping," *Big Data & Society* 5, no. 1 (2018): 1–15.

22. Ensmenger, *Computer Boys Take Over*, 154–155.

23. Paul N. Edwards, *The Closed World: Computers and the Politics of Discourse in Cold War America* (Cambridge, MA: MIT Press), 115, 394n5.

24. Theodora Dryer, "Designing Certainty: The Rise of Algorithmic Computing in an Age of Anxiety 1920–1970" (PhD diss., University of California–San Diego, 2019); Zoë Hitzig, "Optimize This," *ŠUM: Journal for Contemporary Art Criticism* 15 (2020): 2064–2080.

25. Mirowski, *Machine Dreams*; Stephen P. Waring, *Taylorism Transformed: Scientific Management Theory since 1945* (Chapel Hill: University of North Carolina Press, 1991), chap. 2; Marion Fourcade and Rakesh Khurana, "From Social Control to Financial Economics: The Linked Ecologies of Economics and Business in Twentieth Century America," *Theory & Society* 42, no. 2 (2013): 121–159.

26. Herbert A. Simon, "Two Heads Are Better than One: The Collaboration between AI and OR," *Interfaces* 17, no. 4 (1987): 10.

27. Letter and Industrial Dynamics Pamphlet from Jay W. Forrester to Frank M. Verzuh, March 29, 1959, AC.0062, box 2, folder 22, Records of the Massachusetts Institute of Technology Computation Center, MIT Libraries Department of Distinctive Collections.

28. Harlan D. Mills, "Marketing as a Science," *Harvard Business Review* 39, no. 5 (1961): 137.

29. Alfred A. Kuehn and Ralph L. Day, "Probabilistic Models of Consumer Buying Behavior," *Journal of Marketing* 28, no. 4 (1964): 27.

30. William Lazer, "Operations Research and Marketing Science," in *Science in Marketing*, ed. George Schwartz (New York: John Wiley & Sons, 1965), 435.

31. Lazer, 432.

32. Benjamin Lipstein, "Prospects for the Management Sciences in Advertising," *Management Science* 13, no. 2 (1966): B5.

33. Edwards, *Closed World*, 114.

34. Horvath, "Operations Research," 6.

35. Michael Halbert, "Empirical Research in Consumer Shopping and Motivation," in *Proceedings of the Thirty-Seventh National Conference of the American Marketing Association, New York, December 27th–29th 1955* (New York: American Marketing Association, 1955), 15.

36. Simon, "Two Heads," 15.

37. Peter F. Drucker, *The Practice of Management* (New York: Harper & Row, 1954), 370.

38. Waring, *Taylorism Transformed*, 28.

39. "Business Statistics Going PhD," *Business Week*, May 30, 1953, 96.

40. Eugene Ladin, "The Role of the Accountant in Operations Analysis," *Accounting Review* 37, no. 2 (1962): 291.

41. "Business Statistics Going PhD," 96.

42. George E. Kimball, "Some Industrial Applications of Military Operations Research Methods," *Operations Research* 5, no. 2 (1957): 203.

43. Philip Kotler, "Operations Research in Marketing," *Harvard Business Review* 45, no. 1 (1967): 30–34, 37–38, 40, 42, 44, 187–188; Melvin Anshen, "Management Science in Marketing: Status and Prospects," *Management Science* 2, no. 3 (1956): 222–231.

44. National Research Council, *Operations Research with Special Reference to Non-Military Applications: A Brochure* (Washington, DC: National Academies Press, 1951), 3.

45. Charles Kittel, "The Nature and Development of Operations Research," *Science* 105, no. 2719 (1947): 150.

46. Kittel, 152.

47. Russell L. Ackoff, "Some New Statistical Techniques Applicable to Operations Research," *Journal of the Operations Research Society of America* 1, no. 1 (1952): 10–17; Bernard O. Koopman, "The Optimum Distribution of Effort," *Journal of the Operations Research Society of America* 1, no. 2 (1953): 52–63; John F. Magee, "The Effect of Promotional Effort on Sales," *Journal of the Operations Research Society of America* 1, no. 2 (1953): 64–74.

48. Philip M. Morse, "Trends in Operations Research," *Journal of the Operations Research Society of America* 1, no. 4 (1953): 159.

49. Kimball, "Some Industrial Applications of Military Operations Research Methods," 203.

50. Elizabeth A. Richards, "Operations Research or the Scientific Method," *Journal of Marketing* 19, no. 2 (1954): 159–160.

51. Drucker, *Practice of Management*, 366.

52. Russell L. Ackoff, "Comments on Operations Research," *Journal of Marketing* 20, no. 1 (1955): 47–48.

53. John A. Howard, "Operations Research and Market Research," *Journal of Marketing* 20, no. 2 (1955): 143.

54. Kenneth A. Longman, "Marketing Effort—How Productive Is It?," in *Minutes of the Third Meeting of the Operations Research Discussion Group, Friday, April 22, 1960* (New York: Advertising Research Foundation, 1960), 4.

55. National Research Council, *Operations Research*, 4.

56. Horace C. Levinson, "Experiences in Commercial Operations Research," *Journal of the Operations Research Society of America* 1, no. 4 (1953): 234.

57. National Research Council, *Operations Research*, 4.

58. Levinson, "Experiences in Commercial Operations Research," 221–222.

59. See, e.g., Cyril C. Herrmann and John F. Magee, "'Operations Research' for Management," *Harvard Business Review* 31, no. 4 (1953): 100–112.

60. Marcello Vidale quoted in *Minutes of the First Meeting of the Operations Research Discussion Group, October 5, 1959* (New York: Advertising Research Foundation, 1959), 7.

61. Tibor Fabian quoted in *Minutes of the Second Meeting of the Operations Research Discussion Group, February, 16, 1960* (New York: Advertising Research Foundation, 1960), 1.

62. Herrmann and Magee, "'Operations Research' for Management," 101–102.

63. Jerome D. Herniter and John F. Magee, "Customer Behavior as a Markov Process," *Operations Research* 9, no. 1 (1961): 105.

64. Ronald A. Howard, "Comments on the Origin and Application of Markov Decision Processes," *Operations Research* 50, no. 1 (2002): 100–102.

65. Howard, 100.

66. *Minutes of the Second Meeting of the Operations Research Discussion Group*, 10.

67. *Transcript of the Fourth Meeting of the Operations Research Discussion Group* (New York: Advertising Research Foundation, 1960), 37, Charles H. Sandage Papers, 1910–1999, box 11, Speeches, 1955–1964, Transcripts of Operations Research, University of Illinois Archives, Urbana; emphasis added.

68. Herniter and Magee, "Customer Behavior as a Markov Process," 121–122.

69. Richard Maffei, "A Mathematical Model of Brand Switching," in *Operations Research in Advertising: A Summary of Papers Presented to ARF's Operations Research Discussion Group* (New York: Advertising Research Foundation, 1961), 36.

70. Albert Wesley Frey, *Advertising*, 3rd ed. (New York: The Ronald Press Company, 1961), 3.

71. Frey, 13.

72. Frey, 505.

73. "A Profit Yardstick for Advertising," *Business Week*, November 22, 1958, 49.

74. Charles K. Ramond, "Science in Wonderland," *Journal of Advertising Research* 1, no. 1 (September 1960): 32.

75. John F. Magee, "The Problem: Can the Results of the Ad Budget be Scientifically Predicted?," in *Proceedings of the 4th ARF Conference* (New York: Advertising Research Foundation, 1958), 15.

76. Jay W. Forrester, "The Relationship of Advertising to Corporate Management," in *Proceedings of the 4th ARF Conference*, 76.

77. Forrester, 75, 92.

78. Jay W. Forrester, "Advertising: A Problem in Industrial Dynamics," *Harvard Business Review* 37, no. 2 (1959): 102.

79. Forrester, 110; Forrester, "Relationship of Advertising to Corporate Management," 90.

80. John D. C. Little, "Interview by Robert Klein," September 4, 2014, https://www.informs.org/content/download/348559/3217056/file/Robert%20Klein%20Interviews%20John%20D.%20C.%20Little,%20September%204,%202014.pdf.

81. John D. C. Little and Russell L. Ackoff, "How Techniques of Mathematical Analysis Have Been Used to Determine Advertising Budgets and Strategy," in *Proceedings of the 4th ARF Conference*, 19.

82. Little, "Interview by Robert Klein."

83. *Operations Research in Advertising*, iii.

84. *Operations Research in Advertising*, iii.

85. *A Report of the Ninth Meeting of the ARF Operations Research Discussion Group, November 7, 1963* (New York: Advertising Research Foundation, 1963).

86. Roger Barton, *Advertising Agency Operations and Management* (New York: McGraw-Hill, 1955), 52.

87. Barton, 52–53.

88. Robert Vincent Zacher, *Advertising Techniques and Management* (Homewood, IL: R. D. Irwin, 1961), 31.

89. Robert J. Kegerreis, "Marketing Management and the Computer: An Overview of Conflict and Contrast," *Journal of Marketing* 35, no. 1 (1971): 5.

90. Lipstein, "Prospects for the Management Sciences in Advertising," B2.

91. *Minutes of the First Meeting of the Operations Research Discussion Group*, 4.

92. David W. Miller and Martin K. Starr, *Executive Decisions and Operations Research* (New York: Prentice-Hall, 1961), 190–209.

93. Martin Starr, "Interview by Henrique Correa," May 9, 2019, https://www.informs .org/Explore/History-of-O.R.-Excellence/Biographical-Profiles/Starr-Martin-K.

94. *Minutes of the First Meeting of the Operations Research Discussion Group*, 4.

95. David B. Learner and Fred Y. Philips, "Contributions to Marketing," in *Systems and Management Science by Extremal Methods: Research Honoring Abraham Charnes at Age 70*, ed. Fred Young Philips and John James Rousseau (New York: Springer, 1992), 30–31.

96. See, e.g., A. Charnes, W. W. Cooper, J. K. DeVoe, D. B. Learner, and W. Reinecke, "Generation of Approximation of Reach and Distribution of Frequencies," Systems Research Memorandum 171, Technological Institute, Northwestern University, February 14, 1967, https://apps.dtic.mil/sti/pdfs/AD0656905.pdf.

97. Letter to Dr. R. A. Wachsler, June 5, 1968, William W. Cooper Papers, box 26, folder 15, Carnegie Mellon University Archives.

98. "BBDO Advertising Agency—Meeting Minutes, 1965–1968," William W. Cooper Papers, box 26, folder 15, Carnegie Mellon University Archives.

99. Lewis G. Pringle, "The Academy and the Practice: In Principle, Theory and Practice Are Different, but, in Practice, They Never Are," *Marketing Science* 20, no. 4 (2001): 373.

100. Edward Brody, "Marketing Engineering at BBDO," *Interfaces* 31, no. 3 (2001): S76.

101. "Into 1963," January 1, 1963, 10, JWT New Business Records, box 8, Hartman Center; "Thompson a Pioneer in Field; Started Using Computer in '62," *J. Walter Thompson Company News* 24, no. 5 (January 31, 1969), JWT Staff Newsletters, box MN15, Hartman Center.

102. "An Integrated System for Media Planning," March 14, 1968, JWT Information Center Records, box 13, Documents Relating to Media and Media Buying, Hartman Center.

103. "Advertising Management," J. Walter Thompson New York Office (1970), 4–5.

104. John David, *Operational Research and Marketing* (London: J. Walter Thompson Company, 1967), 1, JWT Publications, box IL2, Hartman Center.

105. David, 14.

106. Philip Kotler, "A Design for the Firm's Marketing Nerve Center," *Business Horizons* 9, no. 3 (1966): 63–74; John D. C. Little, "Adaptive Experimentation," in *A Report of the Twelfth Meeting of the ARF Operations Research Discussion Group, April 19, 1965* (New York: Advertising Research Foundation, 1965), 1–27.

107. A. P. Zentler and Dorothy Ryde, "An Optimum Distribution of Publicity Expenditure in a Private Organisation," *Management Science* 2, no. 4 (1956): 337–352.

108. B. Benjamin and J. Maitland, "Operational Research and Advertising: Some Experiments in the Use of Analogies," *Operational Research Quarterly* 9, no. 3 (1958): 207–217.

109. B. Benjamin, W. P. Jolly, and J. Maitland, "Operational Research and Advertising: Theories of Response," *Operations Research Quarterly* 11, no. 4 (1960): 205–218.

110. Simon Broadbent, "Media Planning and Computers by 1970: A Review of the Use of Mathematical Models in Media Planning," *Journal of the Royal Statistical Society Series C* 15, no. 3 (1966): 242–243.

111. Stanley I. Cohen, "The Rise of Management Science in Advertising," *Management Science* 13, no. 2 (1966): B10.

112. John C. Maloney, "Advertising Research and an Emerging Science of Mass Persuasion," *Journalism Quarterly* 41, no. 4 (1964): 518.

113. Maloney, 521.

114. Albert R. Kroeger, "The Computers Move in on Advertising," *Television* 20, no. 6 (1963): 43.

115. Langhoff, "The Setting," 3.

116. "Ule of K&E: Marketing Prophet," *Television* 17, no. 8 (1960): 89.

117. "Ule of K&E," 52.

118. John Nuccio, "Automation—Final Step in a Media Man's Evolution," *Sponsor*, July 13, 1964, 42.

119. Richard F. Reynolds, "They Laughed When I Sat Down to Play with My Dictionary," *Broadcasting*, June 18, 1962, 24.

120. Robert Bodeau, "Advertising Media: An Appraisal," *Journal of Advertising* 5, no. 1 (1976): 7–8.

121. William T. Moran, "Practical Media Decisions and the Computer," *Journal of Marketing* 27, no. 3 (1963): 32.

122. Lazer, "Operations Research and Marketing Science," 439.

123. "Math Penetrates the Social Sciences," *New York Times*, December 16, 1969, 49–51; Jagdish N. Sheth, "The Multivariate Revolution in Marketing Research," *Journal of Marketing* 35, no. 1 (1971): 13–19.

124. Jagdish N. Sheth, "Multivariate Analysis in Marketing," *Journal of Advertising Research* 10, no. 1 (1970): 29.

125. Philip Kotler, "The Use of Mathematical Models in Marketing," *Journal of Marketing* 27, no. 4 (1963): 41.

126. Fred J. Gross, "Critique of OR," *Journal of Marketing* 30, no. 1 (1966): 108.

127. Little and Ackoff, "How Techniques of Mathematical Analysis Have Been Used," 23.

128. Brody, "Marketing Engineering at BBDO," S77.

129. Charles Ramond, "$E = mc^2$," *Television* 18, no. 2 (1961): 59.

130. Dudley M. Ruch, "Management Decision Guided by Quantitative Research," in *A Report of the Seventh Meeting of the ARF Operations Research Discussion Group* (New York: Advertising Research Foundation, 1962), 3.

131. Herbert A. Simon, "Theories of Decision-Making in Economics and Behavioral Science," *American Economic Review* 49, no. 3 (1959): 253–283.

132. Donald C. Marschner, "Theory versus Practice in Allocating Advertising Money," *Journal of Business* 40, no. 3 (1967): 301.

133. Pillsbury's Dudley Ruch quoted in "OR Discussion Group, Thursday Morning Session, June 28, 1962," in *Report of the Seventh Meeting of the ARF Operations Research Discussion Group*, 7.

134. "Longman Warns of Marketing 'Fads,'" *J. Walter Thompson Company News* 18, no. 22 (June 28, 1963): 8, JWT Staff Newsletters, box MN12, Hartman Center.

135. *Minutes of the Second Meeting of the Operations Research Discussion Group*, 21.

136. Michael Halbert quoted in "Discussion of Learner and Godfrey Papers," in *Mathematical Methods of Media Selection: A Report of the Sixth Meeting of the ARF Operations Research Discussion Group, December 18* (New York: Advertising Research Foundation, 1961), 6.

137. A. S. C. Ehrenberg, "Models of Fact: Examples from Marketing," *Management Science* 16, no. 7 (March 1970): 435–445.

138. A. S. C. Ehrenberg, "Laws in Marketing: A Tail-Piece," *Journal of the Royal Statistical Society Series C Applied Statistics* 15, no. 3 (1966): 267.

139. Ehrenberg, "Models of Fact," 339.

140. John F. Magee, "Application of Operations Research to Marketing and Related Management Problems," *Journal of Marketing* 18, no. 4 (1954): 367.

141. Rieder, "Scrutinizing an Algorithmic Technique," 110.

142. Howard, "Operations Research and Market Research," 145.

143. *Minutes of the First Meeting of the Operations Research Discussion Group*, 13.

144. *Minutes of the First Meeting of the Operations Research Discussion Group*, 14.

145. Kenneth A. Longman, "Galileo Galilei's Ghost," *Management Science* 13, no. 12 (1967): B852–B853.

146. "OR Discussion, Thursday Morning Session, June 28, 1962" in *Report of the Seventh Meeting of the ARF Operations Research Discussion Group*, 23.

147. A. Charnes, W. W. Cooper, D. B. Learner, and F. Y. Phillips, "Management Science and Marketing Management," *Journal of Marketing* 49, no. 2 (1985): 96, 102.

148. Ilyse Liffreing, "Confessions of a Data Scientist," *Digiday*, November 19, 2018, https://digiday.com/marketing/confessions-data-scientist-marketers-dont-know -theyre-asking/.

149. A. C. Nielsen Jr., "Marketing Expenditures: The Staff of Product Life," in *Effective Marketing Coordination: Proceedings of the Forty-Fourth National Conference of the American Marketing Association*, ed. George L. Baker Jr. (Chicago: AMA, 1961), 321.

150. Nielsen, 325.

151. Nielsen, 321.

152. Darrell Blaine Lucas and Steuart Henderson Britt, *Measuring Advertising Effectiveness* (New York: McGraw-Hill, 1963), 379–380.

CHAPTER 4

1. "A Media Rule of Thumb Is Ended," *New York Times*, November 16, 1961, 52; "Mathematical Programming for Better Selection of Advertising Media," *Computers and Automation* 10, no. 12 (1961): 12.

2. Wilson quoted in "Mathematical Programming for Better Selection of Advertising Media," 13.

3. Richard P. Jones, "The Quiet Revolution in Media Planning," *Media Decisions* 2, no. 9 (1967): 36.

4. John D. C. Little and Leonard M. Lodish, "MEDIAC: An On-Line Media Selection System" (working paper 298–67, Alfred P. Sloan School of Management, Massachusetts Institute of Technology, November 1967), 2.

5. Douglas B. Brown, "A Practical Procedure for Media Selection," *Journal of Marketing Research* 4, no. 2 (1967): 263.

6. Philip A. Doherty, "A Closer Look at Operations Research," *Journal of Marketing* 27, no. 2 (1963): 65.

7. David Aaker, "Management Science in Marketing: The State of the Art," *Interfaces* 3, no. 4 (1973): 20.

8. Ronald R. Kline, *The Cybernetics Moment: Or Why We Call Our Age the Information Age* (Baltimore: Johns Hopkins University Press, 2015), 136.

9. Robert S. Weinberg, *An Analytical Approach to Advertising Expenditure Strategy* (New York: Association of National Advertisers, 1960), viii.

10. Martin L. Bell, "Leaders in Marketing: Robert Stanley Weinberg," *Journal of Marketing* 35, no. 1 (1971): 49.

11. Weinberg, *Analytical Approach to Advertising Expenditure Strategy*, 6.

12. Weinberg, 89.

13. Gwyn Collins, "Basic Math for Big Problems," *Journal of Advertising Research* 1, no. 1 (1960): 24.

14. "Mathematical Models and Marketing Strategy," *Television* 17, no. 11 (1960): 58.

15. John F. Magee, "The Problem: Can the Results of the Ad Budget be Scientifically Predicted?," *Proceedings of the 4th ARF Conference* (New York: Advertising Research Foundation, 1958): 15.

16. David B. Montgomery and Glen L. Urban, *Management Science in Marketing* (Englewood Cliffs, NJ: Prentice-Hall, 1969), 95.

17. David W. Miller and Martin K. Starr, *Executive Decisions and Operations Research* (Englewood Cliffs, NJ: Prentice-Hall, 1960), 190.

18. Miller and Starr, 192.

19. Miller and Starr, 193.

20. Alec M. Lee, "Decision Rules for Media Scheduling: Static Campaigns," *Operational Research Quarterly* 13, no. 3 (1962): 229.

21. Simon Broadbent, "Media Planning and Computers by 1970: A Review of the Use of Mathematical Models in Media Planning," *Journal of the Royal Statistical Society Series C* 15, no. 3 (1966): 248.

22. Philip Shabecoff, "Research in Revolution," *Sponsor*, January 28, 1963, 33.

23. Ien Ang, *Desperately Seeking the Audience* (New York: Routledge, 1991), 50.

24. Alexander Henderson and Robert Schlaifer, "Mathematical Programming: Better Information for Better Decision Making," *Harvard Business Review* 32, no. 3 (1954): 94.

25. Wroe Alderson, *Marketing Behavior and Executive Action* (Homewood, IL: R. D. Irwin, 1957), 406.

26. Marion Harper Jr., "A New Profession to Aid Management," *Journal of Marketing* 25, no. 3 (1961): 2.

27. Peter Langhoff, "The Setting: Some Non-Metric Observations," in *Models, Measurement, and Marketing*, ed. Peter Langhoff (Englewood Cliffs, NJ: Prentice-Hall, 1965), 9.

28. Wiebe E. Bijker, "How Is Technology Made? That Is the Question!," *Cambridge Journal of Economics* 34, no. 1 (2010): 63–76.

29. Kenneth J. Arrow, "Decision Theory and Operations Research," *Operations Research* 5, no. 6 (1957): 765.

30. Roger Barton, *Advertising Agency Operations and Management* (New York: McGraw-Hill, 1955), 17.

31. Thomas A. Wright, "The Timebuyer: What's His Future?," *Sponsor*, January 20, 1964, 40.

32. Shabecoff, "Research in Revolution," 26.

33. "Buyers Should Know the Big Picture," *Sponsor*, March 30, 1964, 40. ARB stands for American Research Bureau, and SRDS is Standard Rate and Data Service.

34. Shabecoff, "Research in Revolution," 25.

35. Beardsley Graham, "Is TV Management Flying Blind," *Television* 23, no. 1 (1966): 47.

36. Shabecoff, "Research in Revolution," 25.

37. Albert R. Kroeger, "The Computers Move in on Advertising," *Television* 20, no. 6 (1963): 78.

38. "Media Departments—1966," *Sponsor*, October 23, 1961, 26. See also Jones, "Quiet Revolution in Media Planning," 36.

39. William G. White, "Timebuyer of the Seventies: What Will His Job Be?," *Sponsor*, January 27, 1964, 38.

40. White, 41.

41. E. Lawrence Deckinger, "Media Selection Is an Art, Not a Science," *Television* 17, no. 8 (1960): 49.

42. Joseph St. George, "How Practical Is the Media Model?," *Journal of Marketing* 27, no. 3 (1963): 31–32.

43. Robert S. Weinberg, "Management Science and Marketing Strategy," in *Marketing and the Computer*, ed. Wroe Alderson and Stanley J. Shapiro (Englewood Cliffs, NJ: Prentice-Hall, 1963), 106.

44. David Learner, "Mathematical Programming for Better Media Selection," in *Mathematical Methods of Media Selection: A Report of the Sixth Meeting of the ARF Operations Research Discussion Group* (New York: Advertising Research Foundation, 1961); Clark Wilson, "Linear Programming Basics," in *Proceedings of the 8th Annual ARF Conference* (New York: Advertising Research Foundation, 1962), 78–83.

45. "Questions to Clark L. Wilson and Herbert D. Maneloveg," in *Proceedings of the 8th Annual ARF Conference*, 96.

46. "Mathematical Programming for Better Selection of Advertising Media," 14.

47. Broadbent, "Media Planning and Computers by 1970," 242.

48. "Discussion of Learner and Godfrey Papers," in *Mathematical Methods of Media Selection*, 10.

49. "Computers Speed up Tedious Job of Selecting Advertising Media," *Business Week*, November 18, 1961, 49.

50. Quoted in Robert Dow Buzzell, *Mathematical Models and Marketing Management* (Boston: Division of Research, Graduate School of Business Administration, Harvard University, 1964), 95–96.

51. Dennis H. Gensch, "Computer Models in Advertising Media Selection," *Journal of Marketing Research* 5, no. 4 (1968): 415.

52. Aaker, "Management Science in Marketing," 20.

53. Donald E. Sexton Jr., "Microsimulating Consumer Behavior," in *Models of Buyer Behavior: Conceptual, Quantitative, and Empirical*, ed. Jagdish N. Sheth (New York: Harper & Row, 1974), 88.

54. Ralph L. Day, "Linear Programming in Media Selection," *Journal of Advertising Research* 2 (1962): 44.

55. Broadbent, "Media Planning and Computers by 1970," 242.

56. See, e.g., Martin Kenneth Starr, "Management Science and Marketing Science," *Management Science* 10, no. 3 (1964): 563.

57. Stanley Canter quoted in "Discussion of Learner and Godfrey Papers," in *Mathematical Methods of Media Selection*, 5.

58. Harold Weitz, "The Promise of Simulation in Marketing," *Journal of Marketing* 31, no. 3 (1967): 30.

59. Broadbent, "Media Planning and Computers by 1970," 242.

60. Charles B. Weinberg, "Frontiers in Marketing," *Journal of Advertising Research* 9, no. 2 (1969): 59; A. Charnes, W. W. Cooper, J. K. DeVoe, D. B. Learner, and W. Reinecke, "A Goal Programming Model for Media Planning," *Management Science* 14, no. 8 (1968): B423–B430; John D. C. Little and Leonard M. Lodish, "A Media Selection Model and Its Optimization by Dynamic Programming," *Industrial Management Review* 8, no. 1 (1966): 15–24.

61. "Tune Down Gain, Urges Harper," *Broadcasting*, November 19, 1962, 54.

62. Buzzell, *Mathematical Models and Marketing Management*, 97.

63. Learner, "Mathematical Programming for Better Media Selection," 2.

64. "Advertising Enters Age of Computers," *Sponsor*, January 29, 1962, 28–29.

65. "Mathematical Programming for Better Selection of Advertising Media," 21.

66. "Discussion of Learner and Godfrey Papers," 1.

67. Jones, "Quiet Revolution in Media Planning," 38.

68. Stanley I. Cohen, "The Rise of Management Science in Advertising," *Management Science* 13, no. 2 (1966): B25.

69. "Y&R, BBDO Unleash Media Computerization," *Advertising Age*, October 1, 1962, 118.

70. "Questions to William T. Moran, Kenneth Longman (Assistant Director of Research), Joseph F. St. George (Associate Media Director), Young and Rubicam, Inc.," in *Proceedings of the 8th Annual ARF Conference*, 90.

71. Gary L. Lilien and Philip Kotler, *Marketing Decision Making: A Model-Building Approach* (New York: Harper & Row, 1983), 517.

72. "Y&R, BBDO Unleash Media Computerization," 118.

73. Kroeger, "Computers Move in on Advertising," 74.

74. William T. Moran, "Practical Media Decisions and the Computer," *Journal of Marketing* 27, no. 3 (1963): 29–30.

75. Bill Harvey, "A Brief Personal History of Media Optimization" (unpublished manuscript, shared by the author, February 2018), 3.

76. Kroeger, "Computers Move in on Advertising," 74.

77. "Orange-Bordered Reminder Heralds New Golden Age," *J. Walter Thompson Company News* 19, no. 30 (August 14, 1964): 7, JWT Staff Newsletters, box MN 13, Hartman Center.

78. "An Integrated System for Media Planning," March 14, 1968, JWT Information Center Records, box 13, Documents Relating to Media and Media Buying, Hartman Center.

79. "Integrated System for Media Planning," 1.

80. "Integrated System for Media Planning," 2–3.

81. "Integrated System for Media Planning," 3.

82. See, e.g., "Local Market Network/Spot TV Analysis System," January 28, 1972, JWT Staff Newsletters, box DO18, Media Reports, 1972–1973, Hartman Center.

83. "Media Analysis Bibliography 1970–1972," June 1973, JWT Information Center Records, box 13, Documents Relating to Media and Media Buying, Hartman Center.

84. "Advertising Management" (1970), JWT Publications, London Office, box IL3, Hartman Center.

85. Jill Lepore, *If Then: How the Simulmatics Corporation Invented the Future* (New York: Liveright, 2020), 152; "Du Pont Signs to Use Simulmatics' Media-Mix," *Broadcasting*, December 10, 1962, 10.

86. Ithiel de Sola Pool, "Automated New Tool for Decision Makers," *Challenge* 11, no. 6 (1963): 26.

87. "Mechanized 'Dry Runs' for Ad Campaigns," *Broadcasting*, May 28, 1962, 40.

88. "Discussion of Bernstein Paper," in *Mathematical Methods of Media Selection*, 8.

89. Alex Bernstein, "Computer Simulation of Media Exposure," in *Mathematical Methods of Media Selection*, 10.

90. Bernstein, 4.

91. "What Computer Simulation Can Do," *Broadcasting*, October 16, 1961, 38.

92. "Mechanized 'Dry Runs' for Ad Campaigns," 45.

93. Pool, "Automated New Tool," 27.

94. William R. Fair, "Analogue Computations of Business Decisions," *Journal of the Operations Research Society of America* 1, no. 4 (1953): 210–211.

95. Magee, "The Problem," 17.

96. *Transcript of the Fourth Meeting of the Operations Research Discussion Group* (New York: Advertising Research Foundation, 1960), 2.

97. Lepore, *If Then*, 153.

98. Moran, "Practical Media Decisions and the Computer," 30.

99. Moran, 26.

100. "Discussion of Bernstein Paper," 5–6.

101. "Tune Down Gain, Urges Harper," 54.

102. "Canned 'Public'?," *Broadcasting*, April 23, 1962, 44.

103. John R. Hauser and Glen L. Urban, "John D. C. Little," in *Profiles in Operations Research: Pioneers and Innovators*, ed. Arjang A. Assad and Saul I. Gass (New York: Springer, 2011), 662.

104. Franck Cochoy, "Another Discipline for the Market Economy: Marketing as a Performative Knowledge and Know-How for Capitalism," *Sociological Review* 46, no. S1 (1998): 212.

105. John D. C. Little and Leonard Lodish, "A Media Planning Calculus," *Operations Research* 17, no. 1 (1969): 3.

106. Little and Lodish, "MEDIAC," 4.

107. Little and Lodish, 1.

108. Little and Lodish, "Media Planning Calculus," 3.

109. Little and Lodish, "MEDIAC," 6.

110. Little and Lodish, 8.

111. Little and Lodish, "Media Planning Calculus," 31.

112. Little and Lodish, 4.

113. Little and Lodish, 29.

114. Little named Coca-Cola and Nabisco as clients acquired through these seminars. John D. C. Little, "Interview by Robert Klein," September 4, 2014, https://www.informs.org/content/download/348559/3217056/file/Robert%20Klein%20Interviews%20John%20D.%20C.%20Little,%20September%204,%202014.pdf.

115. Philip H. Dougherty, "Computer Aid for Media Men," *New York Times*, June 26, 1968, 70.

116. Montgomery and Urban, *Management Science in Marketing*, 362.

117. Malcom McNiven and Bob D. Hilton, "Reassessing Marketing Information Systems," *Journal of Advertising Research* 10, no. 1 (1970): 6, 9.

118. John D. Leckenby and Heejin Kim, "How Media Directors View Reach/Frequency Estimation: Now and a Decade Ago," *Journal of Advertising Research* 34, no. 5 (1994): 13.

119. Lilien and Kotler *Marketing Decision Making*, 523.

120. Herbert Maneloveg, "A Year of L.P. Media Planning for Clients," in *Proceedings of the 8th Annual ARF Conference*, 88–89.

121. Anna McCarthy, *The Citizen Machine: Governing by Television in 1950s America* (New York: New Press, 2010).

122. "A Profit Yardstick for Advertising," *Business Week*, November 22, 1958, 49–50, 52.

123. Zygmunt Bauman, *Consuming Life* (Cambridge: Polity, 2007).

124. "Questions to Moran, Longman, St. George, Young and Rubicam," 91.

125. "Questions to Wilson and Maneloveg," 97; emphasis added.

126. Philip M. Napoli, *Audience Evolution: New Technologies and the Transformation of Media Audiences* (New York: Columbia University Press, 2011), 30–31.

127. See Bruce G. Carruthers and Wendy Nelson Espeland, "Accounting for Rationality: Double-Entry Bookkeeping and the Rhetoric of Economic Rationality," *American Journal of Sociology* 97, no. 1 (1991): 31–69.

128. Jenna Burrell and Marion Fourcade, "The Society of Algorithms," *Annual Review of Sociology* 47 (2021): 218.

129. Peter Drucker, "Automation—Forerunner of a Marketing Revolution?," in *The Marketing Revolution: Proceedings of the Thirty-Seventh National Conference of the American Marketing Association, New York, December 27th–29th, 1955* (New York: American Marketing Association, 1956), 37.

130. Caitlin Zaloom, *Out of the Pits: Traders and Technology from Chicago to London* (Chicago: University of Chicago Press, 2006), 165.

INTERLUDE

Epigraph: Jay W. Forrester, "Industrial Dynamics," *Harvard Business Review* 36, no. 4 (1958): 37.

1. James J. Gibson, *The Ecological Approach to Visual Perception* (Hillsdale, NJ: Lawrence Erlbaum Associates, 1979); Erica Robles Anderson and Scott Ferguson, "The Visual Cliff: Eleanor Gibson and the Origins of Affordance," *Money on the Left*, April 19, 2022, https://moneyontheleft.org/2022/04/19/the-visual-cliff-eleanor-gibson-the-origins-of-affordance/.

2. Ian Hutchby, "Technologies, Texts and Affordances," *Sociology* 35, no. 2 (2001): 441–456; Paul M. Leonardi, "Theoretical Foundations for the Study of Sociomateriality,"

Information and Organization 23, no. 1 (2013): 59–76; Adrien Shaw, "Encoding and Decoding Affordances: Stuart Hall and Interactive Media Technologies," *Media, Culture & Society* 39, no. 4 (2017): 592–602.

3. Peter Nagy and Gina Neff, "Imagined Affordance: Reconstructing a Keyword for Communication Theory," *Social Media + Society* 1, no. 2 (2015): 1–9.

4. Shoshana Zuboff, "Big Other: Surveillance Capitalism and the Prospects of an Information Civilization," *Journal of Information Technology* 30, no. 1 (2015): 75–89.

5. Carolyn Marvin helped me think through this perspective during many patient conversations.

CHAPTER 5

1. Charles Sinclair, "'Automation Buying' Era Dawns on Agency Row," *Billboard*, September 16, 1957, 1, 13.

2. James W. Cortada, *The Digital Hand*, vol. 2, *How Computers Changed the Work of American Financial, Telecommunications, Media, and Entertainment Industries* (New York: Oxford University Press, 2004), 353.

3. See, e.g., Jeffery F. Rayport, "Is Programmatic Advertising the Future of Marketing?," *Harvard Business Review*, June 22, 2015, https://hbr.org/2015/06/is-programmatic -advertising-the-future-of-marketing.

4. David Beer, "Envisioning the Power of Data Analytics," *Information, Communication & Society* 21, no. 3 (2018): 466.

5. Joseph Turow, *The Daily You: How the New Advertising Industry Is Defining Your Identity and Your Worth* (New Haven, CT: Yale University Press, 2011).

6. Advertising was one of many industries heralding the revolutionary power of computers, management science, and information systems in the 1950s. Thomas Haigh shows how corporations positioned operations research and computers as harbingers of optimization and clerical automation—the same claims discussed in this chapter. Thomas Haigh, "The Chromium-Plated Tabulator: Institutionalizing an Electronic Revolution, 1954–1958," *IEEE Annals of the History of Computing* 23, no. 4 (2001): 75–104; Thomas Haigh, "Inventing Information Systems: The Systems Men and the Computer, 1950–1968," *Business History Review* 75, no. 1 (2001): 15–61.

7. For textbook explanations of spot advertising, see Bruce M. Owen and Steven S. Wildman, *Video Economics* (Cambridge, MA: Harvard University Press, 1999), 11–14; Helen Katz, *The Media Handbook: A Complete Guide to Advertising Media Selection, Planning, Research, and Buying*, 6th ed. (New York: Routledge, 2016), 60.

8. Cathy Taylor, "The Repping of the Web," *Mediaweek* 6, no. 9 (1996): IQ20.

9. "Room at the Top" (advertisement), *Sponsor*, January 9, 1960, 15.

10. "Buyers Should Know the 'Big Picture,'" *Sponsor*, March 30, 1964, 40.

11. "MGM-TV to Ride Demographic Trend," *Sponsor*, December 28, 1964, 15.

12. "JWT Orders Advanced Computer from RCA," *J. Walter Thompson Company News* 17, no. 37 (September 12, 1962): 3, JWT Staff Newsletters, box MN12, Hartman Center.

13. "Walter Herman—The Man behind the Machine," *J. Walter Thompson Company News* 18, no. 3 (January 23, 1963), JWT Staff Newsletters, box MN12, Hartman Center.

14. ARB was bought by CEIR in 1960, and both organizations were absorbed by Control Data Corporation around 1967.

15. "Electronic Brain for Timebuying," *Broadcasting*, May 25, 1959, 31.

16. "Electronic Brain for Timebuying," 31–32.

17. "Automation Buying in Pilot Use at BBDO," *Broadcasting*, November 13, 1961, 9.

18. "Media Buying: With or without the Numbers?," *Broadcasting*, November 20, 1961, 50.

19. Douglas B. Brown and Martin R. Warshaw, "Media Selection by Linear Programming," *Journal of Marketing Research* 2, no. 1 (1965): 86.

20. "Two Big Agencies Pack Electronic Hardware," *Broadcasting*, October 1, 1962, 34.

21. "More Agencies Automate Operations," *Broadcasting*, September 24, 1962, 36.

22. "Tyson's Advice to Time Buyer: 'Gotta Play It the Computer Way,'" *Variety*, January 16, 1963, 50.

23. "Mathematical Programming for Better Selection of Advertising Media," *Computers and Automation* 10, no. 12 (1961): 19.

24. "Two Big Agencies Pack Electronic Hardware," 34.

25. "NAB Convention: Computers Work Praised," *Broadcasting*, April 8, 1963, 66.

26. "Status Report: Agency Automation," *Broadcasting*, July 1, 1963, 32.

27. "Data Processing at H-R in '64," *Broadcasting*, June 17, 1963, 44.

28. "Computer Use Rises," *Sponsor*, April 13, 1964, 40; "A World of Computers Plugged in by Humans," *Broadcasting*, June 1, 1964, 30.

29. "Are Computers Worth What They Cost?," *Broadcasting*, June 13, 1966, 42.

30. "Top 50 Agencies . . . and Their 1966 Radio-TV Billings," *Broadcasting*, November 28, 1966, 30–31.

31. Rena Bartos, *The Future of the Advertising Agency Research Function* (New York: American Association of Advertising Agencies, 1974), 33, JWT Staff Writings & Speeches, box RL.00762, Hartman Center.

32. "Top 50 Agencies . . . and Their 1960 Radio-TV Billings," *Broadcasting*, November 21, 1960, 28–29.

33. "Top 50 Agencies . . . and Their 1966 Radio-TV Billings."

34. "NAB Convention: Computers Work Praised," 66.

35. "Y&R, BBDO Unleash Media Computerization," *Advertising Age*, October 1, 1962, 118.

36. "Are Computers Worth What They Cost?," 43.

37. "Are Computers Worth What They Cost?," 43.

38. Albert R. Kroeger, "The Computers Move in on Advertising," *Television* 20, no. 6 (1963): 45.

39. Simon Broadbent, "Media Planning and Computers by 1970: A Review of the Use of Mathematical Models in Media Planning," *Journal of the Royal Statistical Society Series C* 15, no. 3 (1966): 243.

40. "Y&R, BBDO Unleash Media Computerization," 118.

41. "Are Computers Worth What They Cost?," 43.

42. "Seven Top Agencies to Tackle Computerized Media Planning," *Sponsor*, December 21, 1964, 3.

43. "Are Computers Worth What They Cost?," 43.

44. Richard P. Jones, "The Quiet Revolution in Media Planning," *Media Decisions* 2, no. 9 (1967): 36.

45. Jones, 38.

46. "The Battle of the Computer Marketeers," *Fortune*, January 1, 1965, 171–172.

47. "Advertising: Fine Points Get Finer?," *Broadcasting*, November 6, 1961, 36.

48. A 1963 advertisement showcased this symbiotic relationship, promoting SRDS Data Incorporated—an information service that analyzed masses of media data at "finger-snapping speed"—and Honeywell computers, "the muscle behind this magic." Honeywell is "big in marketing circles," the ad explained: "BBDO, the ad agency, uses a [Honeywell] 400 to help stretch media dollars. A. C. Nielsen Co.— the world's largest market-research organization—has three Honeywell computers." *Wall Street Journal*, June 11, 1963, 30.

49. "The Ad Thinkers and the Computers," *Broadcasting*, October 8, 1962, 36.

50. Richard C. Christian, "The Computer and the Marketing Man," *Journal of Marketing* 26, no. 3 (1962): 80.

51. US Department of Commerce, Bureau of the Census, *Statistical Abstract of the United States, 1970* (Washington, DC: Government Printing Office, 1970), 495.

52. Del DePierro, "TV Buyer's Obstacle Course," *Sponsor*, September 8, 1964, 36.

53. "Reps Surviving Greatest Crisis," *Broadcasting*, February 26, 1962, 80.

54. Erik Barnouw, *The Sponsor: Notes on a Modern Potentate* (New Brunswick, NJ: Transaction Publishers, 2009), 47–58; William Boddy, *Fifties Television: The Industry and Its Critics* (Urbana: University of Illinois Press, 1990), 157–164; Cynthia B. Meyers,

"From Sponsorship to Spots: Advertising and the Development of Electronic Media," in *Media Industries: History, Theory, and Method,* ed. Jennifer Holt and Alisa Perren (Malden, MA: Wiley-Blackwell, 2009), 69–80.

55. Boddy, *Fifties Television,* 159.

56. Sinclair, "'Automation Buying' Era Dawns on Agency Row," 1.

57. Cortada, *Digital Hand,* 2:352–356.

58. "How Data Processing Fits into Networking," *Broadcasting,* June 13, 1966, 48–49.

59. "World of Computers Plugged in by Humans," 34.

60. "Are Computers Worth What They Cost?," 43.

61. "NAB Convention: Computers Work Praised," 66.

62. Gene Accas, "What to Do 'Til the Univac Comes," *Broadcasting,* January 19, 1959, 29.

63. "For Reps, Computers Are the Rainbow's End," *Broadcasting,* June 13, 1966, 52.

64. "Lavin Says TV Rates Are High," *Broadcasting,* November 25, 1963, 57.

65. "There's No Business Like Computer Business," *Broadcasting,* June 27, 1977, 47.

66. "Shake-out among the Station Reps," *Broadcasting,* May 19, 1969, 67; US Department of Commerce, *Statistical Abstract of the United States, 1970,* 497.

67. US Department of Commerce, Bureau of the Census, *Statistical Abstract of the United States, 1980* (Washington, DC: Government Printing Office, 1980), 587.

68. "How TV Business Has Come Back," *Broadcasting,* November 25, 1968, 42–43.

69. "Private Eyes for TV Advertisers," *Broadcasting,* September 11, 1967, 31.

70. George J. Simko, "The Problem of Surviving in the Paperwork Jungle," *Broadcasting,* February 16, 1970, 18.

71. "Commercial Misplacement," *Broadcasting,* July 1, 1963, 5.

72. Media Payment Corporation, "A $149 Misunderstanding" (advertisement), *Broadcasting,* March 26, 1973, 12.

73. Tom Lux, "Computer Power: Taking the Machete to Advertising's Paper Jungle," *Broadcasting,* June 24, 1974, 22.

74. "Computers in the Paper Jungle," *Broadcasting,* June 13, 1966, 42.

75. "Bristol-Myers Hard Sell," *Fortune,* February 1967, 118.

76. "Spot TV's Heaviest Spenders in 1966," *Broadcasting,* April 3, 1967, 56–57.

77. "World of Computers Plugged in by Humans," 32.

78. Robert Heady, "Computers Bring New Dimension to Ad Field," *Advertising Age,* December 10, 1962, 88–90.

79. "Leo Burnett Adopts Computer Processes," *Broadcasting,* October 15, 1962, 30.

80. "Computers to Make Media Buying More Marketing-Oriented, 4A's Told," *Advertising Age*, November 19, 1962, 108.

81. Bruce Hirsch, "Media Department Goes Computerized for Savings in Expense and Time," *Broadcasting*, May 7, 1973, 18.

82. Frank L. McKibbin, "Computerization, the Great Equalizer," *Broadcasting*, May 24, 1971, 18.

83. "Automation Speeds up Spot Sales," *Broadcasting-Telecasting*, October 7, 1957, 70.

84. "Univac to Invade Timebuying, Selling," *Broadcasting-Telecasting*, November 19, 1956, 98.

85. Sinclair, "'Automation Buying' Era Dawns on Agency Row," 2.

86. "Automation Speeds up Spot Sales," 70, 72.

87. "For Reps, Computers Are the Rainbow's End," 52.

88. "Data Processing at H-R in '64," 44.

89. "Data Processing at H-R in '64," 44.

90. "For Reps, Computers Are the Rainbow's End," 50–51.

91. "Codel: The Computer Is Slowed by Doubters," *Broadcasting*, April 25, 1966, 36.

92. "Katz to Get Computer and Increase Staff," *Broadcasting*, November 16, 1964, 34.

93. "There's No Business Like Computer Business," 56.

94. Kroeger, "Computers Move in on Advertising," 69.

95. "Agencies Praise H-R Move," *Broadcasting*, June 17, 1963, 46.

96. See JoAnne Yates, "Business Use of Information and Technology during the Industrial Age," in *A Nation Transformed by Information*, ed. Alfred D. Chandler Jr. and James W. Cortada (New York: Oxford University Press, 2000), 107–135.

97. "JWT Orders Advanced Computer from RCA," *J. Walter Thompson Company News* 17, no. 37 (September 12, 1962): 1, JWT Staff Newsletters, box MN12, Hartman Center.

98. Heady, "Computers Bring New Dimension to Ad Field," 3.

99. Robert Heady, "Computer Taking over Media-Market Selection," *Advertising Age*, December 17, 1962, 25. The source does not stipulate whether this was a one-time or recurring expense.

100. Herb Maneloveg, "The Function of the Media Man: Media Planning" (speech transcript), BBDO Client Media Seminar, St. Moritz Hotel, December 1, 1966, 15.

101. "World of Computers Plugged in by Humans," 30.

102. World of Computers Plugged in by Humans," 32.

103. "Computers Make Media Buying More Marketing-Oriented, 4A's Told," 108.

104. "Shake-out among the Station Reps," 66.

105. "The Competition Gets Keener as the Pie Gets Larger," *Broadcasting*, June 4, 1979, 46.

106. "Tyson's Advice to Time Buyer," 27.

107. Heady, "Computer Taking over Media-Market Selection."

108. "What Food for Computers?," *Broadcasting*, April 29, 1963, 28.

109. Kroeger, "Computers Move in on Advertising," 72.

110. Bennett Cyphers and Gennie Gebhart, "Behind the One-Way Mirror: A Deep Dive into the Technology of Corporate Surveillance," Electronic Frontier Foundation, December 2, 2019, https://www.eff.org/wp/behind-the-one-way-mirror.

111. Simko, "Problem of Surviving in the Paperwork Jungle," 18.

112. George Simko, "New Technology for an 'Old' Problem," *Broadcasting*, November 8, 1971, 14.

113. "Y&R, BBDO Unleash Media Computerization," 118. See also Heady, "Computers Bring New Dimension to Ad Field," 88.

114. "AAAA's Demographic 'White Paper,'" *Broadcasting*, July 15, 1963, 33.

115. "Computers in the Paper Jungle," 42.

116. "Way to Make Spot Buying Easy," *Broadcasting*, September 25, 1967, 31.

117. "How TV Business Has Come Back," *Broadcasting*, November 25, 1968, 43. See also "The Computer and the Buyers and Sellers of of [*sic*] Broadcast Time," *Broadcasting*, April 22, 1974, 44–46.

118. "There's No Business Like Computer Business," 50.

119. Taina Bucher, *If . . . Then: Algorithmic Power and Politics* (New York: Oxford University Press, 2018), 4–5.

120. "SOS Accepted by Most Reps," *Broadcasting*, February 5, 1968, 25.

121. "Agencies Praise H-R Move," 46.

122. "Lavin Says TV Rates are High," 57.

123. "Agencies Praise H-R Move," 46.

124. Malcolm B. Ochs, "Computer Technology: The Potential Revolution in Media Buying," *Broadcasting*, November 27, 1967, 20.

125. "World of Computers Plugged in by Humans," 30

126. "Failures Mark Machine Billing," *Broadcasting*, June 13, 1966, 58.

127. "For Reps, Computers Are the Rainbow's End," 50.

128. "For Reps, Computers Are the Rainbow's End," 50–51.

129. "Future Shock: It's Here Now for the Reps," *Broadcasting*, July 14, 1975, 32.

130. "What Will Television Be Like in '72," *Broadcasting*, November 30, 1967, 39, 42.

131. "ANA Session Spotlights Broadcast Buying," *Broadcasting*, December 10, 1973, 39.

132. "Digital Deal," *Broadcasting*, April 15, 1974, 65.

133. "Instant Avails," *Broadcasting*, January 19, 1976, 7.

134. "World of Computers Plugged in by Humans," 32.

135. "New Setup at Campbell-Ewald," *Broadcasting*, September 6, 1965, 44.

136. "RCA Computer Starts Automation at NH&S," *Broadcasting*, May 16, 1966, 40.

137. Bern Kanner, "The Computer Can Be Useful When Its Limits Are Recognized," *Broadcasting*, October 8, 1962, 26.

138. "Pellegrin Warns of 'Slideruling,'" *Variety*, February 5, 1958, 50.

139. "Should Agencies Get 20% for Spot TV?," *Broadcasting*, April 28, 1968, 20. The consultant was Lee Rich, who became a celebrated television producer.

140. "Shake-out among the Station Reps," 66.

141. "For Reps, Computers Are the Rainbow's End," 51.

142. "Goose and Gander," *Broadcasting*, April 8, 1991, 8.

143. "There's No Business Like Computer Business," 49.

144. "Cosmos Broadcasting Corporation" (advertisement), *Broadcasting*, January 1, 1979, 20–21.

145. Heady, "Computers Bring New Dimension to Ad Field," 88.

146. "Let's Unshackle the Timebuyer," *Broadcasting*, October 24, 1955, 52.

147. "Media Buying: With or without the Numbers?," 56.

148. "Advertising Enters the Age of Computers," *Sponsor*, January 29, 1962, 50.

149. "Electronic Brain for Timebuying," 31.

150. "Buying by Computer," *Broadcasting*, August 12, 1963, 30.

151. Bill Greeley, "Automation Displacing Media Buyer?," *Variety*, February 28, 1962, 34; "People Not Computers, Make Agency Buying Decisions," *Broadcasting*, October 5, 1964, 40.

152. "Ad Thinkers and the Computers," 36.

153. William A. Murphy, "Of Timebuyers: Human and Machine," *Sponsor*, January 13, 1964, 40.

154. John Nuccio, "Automation—Final Step in a Media Man's Evolution," *Sponsor*, July 13, 1964, 43.

155. "Media Departments—1966," *Sponsor*, October 23, 1961, 25–30; "Computers' Future," *Broadcasting*, January 14, 1963, 43–44; "NAB Convention: Computers Work Praised," 66, 68; Lydia R. Reeve, "Challenge: Selling Time in the Computer Age," *Broadcasting*, April 4, 1966, 32. Describing programmatic buying in 2014, the vice

president of US investment at Interpublic's MagnaGlobal media agency said, "This whole thing is about freeing up those [human] resources for more actual human conversation and human creativity." Tim Peterson and Alex Kantrowitz, "The CMO's Guide to Programmatic Buying," *Advertising Age*, May 19, 2014, 25.

156. "Ross Roy Spot Buying Speeded by GE Computer," *Broadcasting*, December 15, 1969, 42; "EDP for Ad Agencies," *Broadcasting*, April 16, 1962, 40; "Computers to Make Media Buying More Marketing-Oriented, 4A's Told," 108; Kroeger, "Computers Move in on Advertising," 42.

157. "Advertising Enters the Age of Computers," 50.

158. Ithiel de Sola Pool, "Automation: New Tool for Decision Makers," *Challenge* 11, no. 6 (1963): 26.

159. Lynn Spigel, *Make Room for TV: Television and the Family Ideal in Postwar America* (Chicago: University of Chicago Press, 1992).

160. "Y&R, BBDO Unleash Media Computerization," 118.

161. Thomas Haigh, "We Have Never Been Digital," *Communications of the ACM* 57, no. 9 (2014): 24.

162. Heady, "Computer Taking over Media-Market Selection," 3.

163. "Shake-out among the Station Reps," 66.

164. "New Middlemen in Spot," *Broadcasting*, July 15, 1968, 23–24; "Middleman— The Specialist of the Future?," *Broadcasting*, September 29, 1969, 33–35.

165. Leo Bogart, "Buying Services and the Media Marketplace," *Journal of Advertising Research* 40, no. 5 (2000): 37–41.

166. Kelly O'Neill and Paul Schulman, "Goodbye to Glad Hand, Farewell to Free Lunches," *Broadcasting*, May 24, 1976, 14.

167. Helen Katz, "How Major U.S. Advertising Agencies Are Coping with Data Overload," *Journal of Advertising Research* 31, no. 1 (1991): 16.

168. "Media Systems Catalogue," memorandum, July 31, 1989, JWT Chicago Office Media Resources and Research Department Records, box 4, Hartman Center.

CHAPTER 6

1. Margaret Graham, "The Threshold of the Information Age: Radio, Television, and Motion Pictures Mobilize the Nation," in *A Nation Transformed by Information: How Information Has Shaped the United States from Colonial Times to the Present*, ed. Alfred D. Chandler Jr. and James W. Cortada (New York: Oxford University Press, 2000), 138.

2. Segmentation, in general, is even older. Ronald A. Fullerton, "Segmentation in Practice: An Historical Overview of the Eighteenth and Nineteenth Centuries," in

The Routledge Companion to Marketing History, ed. D. G. Brian Jones and Marj Tadajewski (New York: Routledge, 2016), 85–95.

3. John F. Lawrence, "Many Big Firms Revise Advertising Programs, Work to Trim Waste," *Wall Street Journal*, May 28, 1963, 1, 16.

4. Quoted in Leo Bogart, "Is It Time to Discard the Audience Concept?," *Journal of Marketing* 30, no. 1 (1966): 52.

5. Jennifer Karns Alexander, *The Mantra of Efficiency: From Waterwheel to Social Control* (Baltimore: Johns Hopkins University Press, 2008).

6. "Computers Open New Vistas in Pinpointing Audiences, Seminar Told," *Advertising Age*, May 26, 1964, 3, 98.

7. "Two Big Agencies Pack Electronic Hardware," *Broadcasting*, October 1, 1962, 34; emphasis added.

8. Frank L. McKibbin, "Computerization, the Great Equalizer," *Broadcasting*, May 24, 1971, 18.

9. Kelly O'Neill and Paul Schulman, "Goodbye to Glad Hand, Farewell to Free Lunches," *Broadcasting*, May 24, 1976, 14.

10. *Are There Consumer Types?* (New York: Advertising Research Foundation, 1964).

11. Albert R. Kroeger, "The Computers Move In on Advertising," *Television*, June 1963, 43.

12. "Computers to Make Media Buying More Marketing-Oriented, 4A's Told," *Advertising Age*, November 19, 1962, 3.

13. "Advertising Management" (1970), JWT Publications, London Office, box IL3, Hartman Center.

14. Philip M. Napoli, *Social Media and the Public Interest: Media Regulation in the Disinformation Age* (New York: Columbia University Press, 2019), 22–40.

15. *Cable Avails*, June–July 1999.

16. Local origination often meant a channel displaying time, temperature, and advertisements printed on small cue cards, orchestrated by a clever but somewhat ramshackle mechanism that automatically rotated a camera across a clock, a thermometer, and one or more ads positioned about a foot in front of the camera. One humorous anecdote I heard involves a cable system in Hawaii that received agitated calls from viewers about a dragon-like monster terrorizing the local channel. In fact, the light affixed to the apparatus displaying the time and temperature had attracted moths, which in turn had attracted a lizard. The reptile appeared quite monstrous crawling only inches from the camera lens. This story was related to me at the Cable Center's Barco Library in Denver.

17. "CATV Headed for Ad-Supported Network?," *Broadcasting*, May 4, 1970, 23–24.

18. Bruce B. Cox, "Cable Advertising: Where to Buy It," *Broadcasting*, January 27, 1975, 14.

19. Cox, 14.

20. Cox, 14.

21. Interview with Jim Chiddix (former chief technical officer, Time Warner Cable; former president of Interactive Video Group, Time Warner Cable; former CEO and chairman, OpenTV), January 28, 2018.

22. Texscan, *Cable Advertising: The First Comprehensive Guidebook for System Operators* (1980), 8, Barco Library, Cable Center, University of Denver (hereafter Barco Library).

23. Texscan, 4; Patrick R. Parsons and Robert M. Frieden, *The Cable and Satellite Television Industries* (Boston: Allyn & Bacon, 1998), 212–213.

24. Chuck Moozakis, "Return on Your Ad Sales Investment," *Cable Television Business*, April 15, 1983, 36.

25. Interview with Larry Zipin (former corporate vice president of advertising sales, Time Warner Cable), January 26, 2018.

26. "Here Comes Another Quantum Leap," *Broadcasting*, November 15, 1982, 68.

27. Paul E. Oliver, "Local Commercial Insertion: A Partnership Cable Operator, Programming Service and Manufacturer," *NCTA Technical Papers* (1983): 1–5.

28. Charles Kadlec, "Problems and Potentials of Cable TV Advertising," *Broadcasting*, September 3, 1984, 16.

29. Zipin interview.

30. "Addressing Addressability," *Broadcasting*, June 8, 1981, 85.

31. "Zenith Z-TAC" (advertisement), *Cable Television Business*, April 15, 1983, 24.

32. "Applications of Addressability Examined by Panel," *Broadcasting*, May 10, 1982, 75, 78.

33. Larry C. Brown, "Addressable Control—A Big First Step toward the Marriage of Computer, Cable, and Consumer," *NCTA Technical Papers* (1981): 42–46.

34. "The Righteous Wrath of Jack Valenti," *Broadcasting*, February 14, 1983, 62.

35. Patrick R. Parsons, *Blue Skies: A History of Cable Television* (Philadelphia: Temple University Press, 2008), 551.

36. Ross Benes, "Linear Addressable TV Ad Spend Will Grow 33.1% This Year," *eMarketer*, May 11, 2021, https://www.emarketer.com/content/linear-addressable-tv-ad-spending-will-grow-33-percent-this-year.

37. "On the Brink with Cable TV," *Broadcasting*, July 5, 1971, 16; "The Gear for Pay, Two-Way Service Stars at NCTA," *Broadcasting*, June 25, 1973, 39–40.

38. Brown, "Addressable Control," 42. Pioneer manufactured the set-top boxes that supported Warner's futuristic Qube system. Qube provided a template for a wide

range of interactive applications, as well as incubating the program networks that matured into MTV and Nickelodeon.

39. Scientific-Atlanta Inc., "8500 Addressable Terminal" (brochure), April 1982, Barco Library.

40. Paul Baran, "Packetcable: A New Interactive Cable System Technology," *NCTA Technical Papers* (1982): 1.

41. SNL Kagan, "U.S. Multichannel Industry Benchmarks," SNL Media & Telecommunications dataset, accessed September 1, 2017, https://www.marketplace.spglobal.com/en/datasets/snl-media-telecommunications-(194).

42. "Sure, Boss!," *Cable Avails*, June 1993, 16.

43. Gary Kim, "New Gear, New Markets Promise Higher Ad Gains," *Multichannel News*, September 24, 1990, 58; "By the Numbers," *Cable Avails*, September 1995, 20.

44. "Local Cable Requires Major Overhaul," *Advertising Age*, April 14, 1997, S9.

45. "The Hot Market in Local Cable Ad Sales," *Broadcasting & Cable*, June 24, 1996, 58.

46. "Sure, Boss!," 16.

47. "Machinova Named to Top TCI Post," *Cable Avails*, February 1995, 1.

48. David E. Wachob (General Instrument Corporation), "Method and Apparatus for Providing Demographically Targeted Television Commercials," US Patent Number 5,155,591, October 13, 1992.

49. J. Scott Bachman and Stephen Dukes, "Compressed Digital Commercial Insertion: New Technology Architectures for the Cable Advertising Business," *NCTA Technical Papers* (1992): 447.

50. Dave Zornow, "Multimedia: Who Pays?," *Cable Avails*, June 1993, 23.

51. Jim McConville, "Direct-Mail Cable on Tap," *Broadcasting & Cable*, December 16, 1996, 88.

52. Rich Brown, "Cable Moving to Digital Ad Inserts," *Broadcasting*, April 6, 1992, 52–53.

53. "Exhorting Cable Advertising's Faithful," *Broadcasting*, April 2, 1984, 62, 64.

54. "Cable TV: An Advertising Medium Coming of Age," *Broadcasting*, April 18, 1988, 87.

55. See, e.g., this exchange between a sales executive at ABC-TV and the vice president of research at Warner Amex Satellite Entertainment Company: Walter Flynn, "A Broadcaster's View of Cable Research," *Broadcasting*, April 12, 1982, 26; Marshall Cohen, "Open Mike: He Who Hesitates . . . ," *Broadcasting*, May 31, 1982, 19.

56. "Open Mike: Computer Solution," *Broadcasting*, June 14, 1982, 22, 26.

57. "Pioneer M5P Addressable Controller System" (brochure, 1993), Barco Library.

58. Pioneer, "The BA-6000" (brochure, 1988), Barco Library.

59. Wachob, "Method and Apparatus for Providing Demographically Targeted Television Commercials."

60. "Changing Nature of Cable Advertising," *Broadcasting*, April 15, 1991, 78.

61. Michael J. Freeman (ACTV Inc.), "Method for Providing Targeted Profile Interactive CATV Displays," US Patent Number 4,602,279, July 22, 1986.

62. "More MSO's Want into Rep Business," *Broadcasting*, April 18, 1988, 89.

63. "Changing Nature of Cable Advertising," 79.

64. Interview with Bruce Anderson (chief operations officer and global chief technology officer, Invidi Technologies Corporation), February 20, 2018.

65. Interview with Chet Kanojia (former CEO, Navic Networks), February 12, 2018.

66. "Cable Advertising Approaching $2 Billion," *Broadcasting*, April 10, 1989, 49–50; Michael Katz, "Local Cable's Ace in the Hole: Digital Interconnection," *Broadcasting & Cable*, June 24, 1996, 59.

67. Interview with Paul Woidke (former chief technology officer, Adlink; former senior vice president of technology, Comcast Spotlight), February 9, 2018.

68. Jill Marks, "Power Bases: Clustering for Numbers," *Cable Television Business*, April 15, 1983, 18, 25.

69. "Exhorting Cable Advertising's Faithful," 62, 64.

70. "At Large," *Broadcasting*, November 8, 1982, 48, 50, 52, 54, 58; "Group W Presents Picture of Confidence to Security Analysts," *Broadcasting*, June 7, 1982, 68, 70, 73. See also Megan Mullen, *Television in the Multichannel Age* (Malden, MA: Blackwell Publishing, 2008), 144–146.

71. "Exhorting Cable Advertising's Faithful," 62, 64; "Nationwide," *Broadcasting*, February 24, 1986, 12.

72. "Sawhill Takes over at NCC," *Broadcasting & Cable*, December 9, 1996, 105.

73. Chuck Ross, "TCI Buys Equity Stake in Cable TV Rep NCC," *Advertising Age*, September 28, 1998, 68.

74. "Leading National Cable Reps Mulling Merger," *Advertising Age*, July 13, 1998, 3.

75. "Cable Advertising Approaching $2 Billion," *Broadcasting*, April 10, 1989, 49–50.

76. Zipin interview.

77. "Satellite Cable Interconnect," *Broadcasting*, September 19, 1988, 77.

78. Katz, "Local Cable's Ace in the Hole," 58.

79. Woidke interview.

80. Alan Salomon, "Cable's Mass Customization a Hit with Dealers," *Advertising Age*, April 12, 1999, 16.

81. Anderson interview.

82. Hernan Galperin, *New Media, Old Politics: The Transition to Digital TV in the United States and Britain* (Cambridge: Cambridge University Press, 2004).

83. Bachman and Dukes, "Compressed Digital Commercial Insertion," 447.

84. Bachman and Dukes, 448; emphasis added.

85. Bachman and Dukes, 453.

86. "Digital Compression: Cable Insertion's Enabling Technology," *Broadcasting*, April 6, 1992, 52–53.

87. "Digital Ads," *Broadcasting*, January 20, 1992, 41. See also Gary Kim, "Industry Ad Execs See Digital Future," *Multichannel News*, February 3, 1992, 28; Brown, "Cable Moving to Digital Ad Inserts," 52–53; "Digital Compression," 52–53.

88. "Digital Advertising," *Cable Week* [TV show], episode 36, August 20, 1993, Barco Library.

89. "Digital Compression," 53.

90. "Digital Advertising."

91. "Digital Insertion Vendors Hit CAB on a Roll," *Broadcasting & Cable*, June 24, 1996, 60. See also Chris McConnell, "Digital Equipment Eyes Ad Insertion Market," *Broadcasting & Cable*, February 6, 1995, 56; "Chicago to Launch Digital Ad Insertion System," *Broadcasting & Cable*, November 20, 1995, 42, 44; "Detroit to Build Fiber Interconnect," *Broadcasting & Cable*, October 7, 1996, 92.

92. "Digital Insertion Vendors Hit CAB on a Roll," 60.

93. Katz, "Local Cable's Ace in the Hole," 58.

94. Chiddix interview.

95. Glen Dickson, "Digital Goes to Work for Adlink," *Broadcasting & Cable*, November 6, 1996, 108.

96. James Dunaway, "Los Angeles," *Mediaweek*, October 25, 1999, 22–28.

97. "Adlink" (advertisement), *Broadcasting & Cable*, October 26, 1998, 47.

98. "The New Interconnects," *Broadcasting & Cable*, November 23, 1998, 38.

99. "Adlink Targets New Brand Campaign," *Cable Avails*, November 1998, 6.

100. Joan Van Tassel, "Cable's Trump Card: Digital Ad Insertion," *Broadcasting & Cable*, April 14, 1997, 76.

101. Woidke interview.

102. "Traffic and Billing Goes to Top of MSO's List," *Cable Avails*, December 1995, 9.

103. Zipin interview.

104. McConville, "Direct-Mail Cable on Tap."

105. "More Machinations in Traffic and Billing," *Cable Avails*, June–July 1999, 13.

106. "Big-Time Breakout Boosted in Buying Local Cable Time," *Advertising Age*, November 30, 1998, S7, S13.

107. Zipin interview.

108. "NCC Advancing EDI after TCI Deal," *Cable Avails*, November 1998, 8.

109. "Big-Time Breakout Boosted in Buying Local Cable Time," S7, S13.

110. "Bargains May Be over for T/B Systems," *Cable Avails*, February 1995, 27.

111. Kevin Bowe, "Thinking All the Way through the EDI Proposition," *Cable Avails*, June–July 1999, 28.

112. Kanojia interview.

113. William Boddy, *New Media and Popular Imagination: Launching Radio, Television, and Digital Media in the United States* (New York: Oxford University Press, 2004); Joseph Turow, *Niche Envy: Marketing Discrimination in the Digital Age* (Cambridge, MA: MIT Press, 2008).

114. Joe Mandese, "Death Knell for Demo?," *Advertising Age*, July 25, 1994, S2.

115. Interview with Bruce Thomas (former vice president of national ad sales, Tele-Communications Inc.; former general manager, Comcast Spotlight), February 2, 2018.

116. L. J. Davis, *The Billionaire Shell Game* (New York: Doubleday, 1998).

117. Anderson interview.

118. Tobi Elkin, "Addressable Ad Trials Start Up," *Advertising Age*, December 3, 2001, 42. Comcast soon acquired AT&T Broadband and most of the former TCI properties.

CHAPTER 7

Epigraph: George Winslow, "Getting TV Ads to Click with Consumers," *Broadcasting & Cable*, February 20, 2012, 27.

1. Thomas Streeter, "The Cable Fable Revisited: Discourse, Policy, and the Making of Cable Television," *Critical Studies in Mass Communication* 4, no. 2 (1987): 174–200.

2. Patrick R. Parsons, *Blue Skies: A History of Cable Television* (Philadelphia: Temple University Press, 2008), ix.

3. Robert W. McChesney, John Bellamy Foster, Inger Stole, and Hannah Holleman, "The Sales Effort and Monopoly Capital," *Monthly Review* 60, no. 11 (2009): 15.

4. For more on transactivity, see Darin Barney, *Prometheus Wired: The Hope for Democracy in the Age of Network Technology* (Vancouver, BC: UBC Press, 2000), 163–167.

5. Emily Hund and Lee McGuigan, "A Shoppable Life: Performance, Selfhood, and Influence in the Social Media Storefront," *Communication, Culture & Critique* 12, no. 1 (2019): 18–35.

6. Carolyn Marvin, *When Old Technologies Were New: Thinking about Electric Communication in the Late Nineteenth Century* (New York: Oxford University Press, 1988).

7. Peter Nagy and Gina Neff, "Imagined Affordance: Reconstructing a Keyword for Communication Theory," *Social Media + Society* 1, no. 2 (2015): 1–9.

8. Mads Borup, Nik Brown, Kornella Konrad, and Harro van Lente, "The Sociology of Expectations in Science and Technology," *Technology Analysis & Strategic Management* 18, nos. 3–4 (2006): 285–298; Neil Pollock and Robin Williams, "The Business of Expectations: How Promissory Organizations Shape Technology and Innovation," *Social Studies of Science* 40, no. 4 (2010): 525–548.

9. For exemplary studies of these dynamics, see Susan J. Douglas, *Inventing American Broadcasting, 1899–1922* (Baltimore: Johns Hopkins University Press, 1987); Donald MacKenzie, *Inventing Accuracy: A Historical Sociology of Nuclear Missile Guidance* (Cambridge, MA: MIT Press, 1990).

10. Tarleton Gillespie, "The Politics of 'Platforms,'" *New Media & Society* 12, no. 3 (2010): 347–364.

11. Jonathan Gray, *Television Entertainment* (New York: Routledge, 2008), 122.

12. Raymond Williams, *Television: Technology and Cultural Form*, rev. ed. (New York: Routledge, 2003), 114.

13. Bill Gates, *The Road Ahead* (New York: Viking Penguin, 1995), 165.

14. Garth Johnston, "The iTV Evangelist," *Broadcasting & Cable*, March 19, 2007, 32.

15. Josh Bernoff, *Smarter Television* (Cambridge, MA: Forrester Research, 2000), 13.

16. Bernoff, 6–8.

17. Johnston, "iTV Evangelist."

18. "Clicking on Jennifer Aniston's Sweater," TV of Tomorrow Show, San Francisco, March 11, 2008; "Tcommerce: Transforming the Economics of Television," TV of Tomorrow Show, New York City, November 17, 2010; "The Changing Economics of Basic Cable Networks in the OTT Era," SNL Kagan's 6th Annual Multichannel Summit, New York City, November 19, 2015; "From CPM to ROI: The Science of Attribution and Its Impact on TV Advertising," TV of Tomorrow Show (virtual conference), April 27, 2021.

19. Jon Lafayette, "Samsung Adds T-Commerce Apps to Smart Sets," *Broadcasting & Cable*, January 7, 2013, http://www.broadcastingcable.com/news/technology/ces -samsung-adds-t-commerce-apps-smart-sets/49719.

20. Susan Strasser, *Satisfaction Guaranteed: The Making of the American Mass Market* (New York: Pantheon Books, 1989), 24; Richard R. John, *Network Nation: Inventing American Telecommunications* (Cambridge, MA: Belknap Press of Harvard University, 2010), 289.

21. Conversely, telephones also permitted unwelcome strangers to intrude into the home. Marvin, *When Old Technologies Were New*, chap. 2.

22. Bill Bailey, "P&G, Biggest Air User Doubles Sales," *Broadcasting*, June 4, 1945, 18, 28.

23. Noah Arceneaux, "A Sales Floor in the Sky: Philadelphia Department Stores and the Radio Boom of the 1920s," *Journal of Broadcasting & Electronic Media* 53, no. 1 (2009): 76–89.

24. William Leach, *Land of Desire: Merchants, Power, and the Rise of a New American Culture* (New York: Vintage, 1994).

25. "Lapel Mike for Department Stores," *Broadcasting*, May 1, 1933, 14.

26. "Macy's Going to Try WABD for Retail Selling," *Billboard*, December 30, 1944, 7. For more on WABD's television department store and "armchair shopping" strategy, see Lynn Spigel, *Make Room for TV: Television and the Family Ideal in Postwar America* (Chicago: University of Chicago Press, 1992), 79.

27. "Sell-a-Vision," *Broadcasting-Telecasting*, December 12, 1949, 64.

28. Spigel, *Make Room for TV*, 82–83.

29. "The Cox Plan for Cable TV," *Broadcasting*, November 25, 1968, 54; Nicholas Johnson, "CATV," *Saturday Review*, November 11, 1967, 87–88, 96–98.

30. Ralph Lee Smith, "The Wired Nation," *Nation*, May 18, 1970, 582–606; Eileen R. Meehan, "Technical Capability versus Corporate Imperatives: Toward a Political Economy of Cable Television and Information Diversity," in *The Political Economy of Information*, ed. Vincent Mosco and Janet Wasko (Madison: University of Wisconsin Press, 1988), 167–187.

31. Quoted in Parsons, *Blue Skies*, 239.

32. This discussion draws on Williams's argument that the institution of commercial broadcasting was produced within a contradictory social tendency toward both dislocation and home-centered living, which he called mobile privatization. *Television*, 19–22.

33. Parsons, *Blue Skies*, 233.

34. Smith, "Wired Nation," 584.

35. Parsons, *Blue Skies*, 259–268.

36. "Cable Promoters Start Thinking Big," *Broadcasting*, January 26, 1970, 54–57; "On the Brink with Cable TV," *Broadcasting*, July 5, 1971, 18; "How High Is the Peak on Cable's Mountain?," *Broadcasting*, January 2, 1978, 41, 44–45.

37. "Show Your Stuff, Burch Tells NCTA," *Broadcasting*, July 12, 1971, 18–26.

38. Walter S. Baer, *Interactive Television: Prospects for Two-Way Services on Cable* (Santa Monica, CA: Rand Corporation, 1971), 39.

39. "Kahn Plans a Satellite for CATV," *Broadcasting*, February 9, 1970, 38.

40. "Kahn Plans a Satellite for CATV."

41. "On the Brink with Cable TV," 19.

42. "Sides Drawn on N.Y. Cable Plans," *Broadcasting*, July 27, 1970, 50–51.

43. "Dialing up a Dialogue on Cable," *Broadcasting*, May 31, 1971, 45.

44. "Where Pay Cable Is Getting Started," *Broadcasting*, April 9, 1973, 47–49.

45. Erik Barnouw, *Tube of Plenty*, 2nd rev. ed. (New York: Oxford University Press, 1990), 498. For example, a game show called *Majic Touch* let viewers purchase merchandise from its sponsor, Lazarus department stores. "Horatio Alger Advice to Cable Advertising," *Broadcasting*, February 15, 1981, 66.

46. "Schmidt Re-emerged with New Company in Cable World," *Broadcasting*, December 3, 1979, 64–65.

47. L. J. Davis, *The Billionaire Shell Game* (New York: Doubleday, 1998), 54.

48. "Cox Cable Gets Omaha Franchise," *Broadcasting*, August 25, 1980, 110–111; "Cox Gets Another," *Broadcasting*, January 5, 1981, 34.

49. "ATC-Daniels Partnership Takes Denver Cable Franchise," *Broadcasting*, March 1, 1982, 34.

50. "They're on the Line in Chicago," *Broadcasting*, September 6, 1982, 28–29.

51. "Franchise Promises Come Back to Haunt Warner," *Broadcasting*, January 16, 1984, 39–40.

52. Barnouw, *Tube of Plenty*, 500.

53. "Showdown in Cerritos: Telco, Cable at Odds over System Control," *Broadcasting*, January 4, 1988, 45–46.

54. See Parsons, *Blue Skies*, 600–608, 643–647, 686–688.

55. "GTE Details Fiber Plans for Cerritos," *Broadcasting*, June 27, 1988, 66, 68; "Widening Main Street," *Broadcasting*, December 12, 1988, 10.

56. Mark Bernicker, "Independent Telcos Moving Quickly to Provide Video Services," *Broadcasting & Cable*, May 2, 1994, 41.

57. "TeleMediaWatch," *Broadcasting & Cable*, January 31, 1994, 14.

58. Mark Berniker, "US West Building an Interactive Mall," *Broadcasting & Cable*, August 1, 1994, 34, 36.

59. Dennis Wharton, "FCC Gives Bell Atlantic OK to Go Hollywood," *Variety*, January 16–22, 1995, 38.

60. Harry A. Jessell, "VOD, Gaming to Help Pave Superhighway," *Broadcasting & Cable*, December 6, 1993, 6, 10.

61. "Ellison: TV's Oracle," *Broadcasting & Cable*, January 17, 1994, 86.

62. Jim McConville, "News, Sports, Pizza from Time Warner," *Broadcasting & Cable*, December 4, 1995, 83.

63. "Cox's Jim Kennedy," *Broadcasting & Cable*, June 20, 1994, 23–27.

64. Price Colman, "Stand and Deliver," *Broadcasting & Cable*, December 8, 1997, 42.

65. "The Gear for Pay, Two-Way Services Stars at NCTA," *Broadcasting*, June 25, 1973, 39–40.

66. "Interactivity Seen as Key to Future Growth," *Broadcasting*, November 25, 1991, 37.

67. General Instrument, "CFT 2200" (product brochure, 1996), Barco Library.

68. Price Colman, "CableLabs in the Middle on Set-Tops," *Broadcasting & Cable*, October 20, 1997, 56.

69. William Boddy, *New Media and Popular Imagination: Launching Radio, Television, and Digital Media in the United States* (New York: Oxford University Press, 2004), 137–143; Dan Schiller, *Digital Capitalism: Networking the Global Market System* (Cambridge, MA: MIT Press, 1999), 108–113.

70. General Instrument and Scientific-Atlanta were both bought by larger hardware manufacturers—GI by Motorola in 1999 and S-A by Cisco in 2005. Google later acquired Motorola's STB business before appropriating the company's patents and selling the business again.

71. Andrew Kupfer and Byron Harmon, "Set-Top Box Wars," *Fortune*, August 22, 1994, 110–116; Price Colman, "Set-Top Scramble in Silicon Valley," *Broadcasting & Cable*, December 8, 1997, 14.

72. Catherine M. Skelly and Matthew Weiss, *Turning TV Sets into Cash Registers* (New York: Gruntal, 2000), 101.

73. Skelly and Weiss, 101.

74. Sean Scully, "Preparing to Convert the Converter." *Broadcasting & Cable*, May 3, 1993, 59.

75. Price Colman, "Interactive TV Ads on the Horizon," *Broadcasting & Cable*, January 4, 1999, 66.

76. Colman, 66.

77. Schiller, *Digital Capitalism*, 109.

78. Devin Leonard, "The Most Valuable Square Foot in America," *Fortune*, April 1, 2002.

79. "Revenues for Interactive TV Shopping to Reach $4.3 Billion by 2005," *Direct Marketing* 64, no. 5 (2001): 8.

80. Skelly and Weiss, *Turning TV Sets into Cash Registers*, 38–43.

81. Mark Berniker, "Set-Top Chaos," *Broadcasting & Cable*, September 25, 1995, 58.

82. Mark Berniker, "Full Service Network Starts Small," *Broadcasting & Cable*, December 19, 1994, 18.

83. Mark Berniker, "Telcos Seek $800 Million Worth of Set-Top Boxes," *Broadcasting & Cable*, March 6, 1995, 31, 34.

84. Hernan Galperin and François Bar, "The Regulation of Interactive Television in the United States and the European Union," *Federal Communications Law Journal* 55, no. 1 (2002): 61–84.

85. Leonard, "Most Valuable Square Foot in America."

86. Rich Brown and Richard Tedesco, "Promise vs. Performance," *Broadcasting & Cable*, April 29, 1996, 10, 12.

87. Craig Leddy, "Peering behind Jennifer Aniston's Sweater," *Multichannel News*, February 26, 2001, 52.

88. Monica Hogan, "Convergence Draws Cable to T-Commerce," *Multichannel News*, May 15, 2000, 76.

89. R. Cole, "Maybe There Isn't Gold in Rachel's Sweater," *Cable World*, June 18, 2001, 25.

90. Stuart Elliot, "Remote Clicks that Do More than Just Change Channels," *New York Times*, June 16, 2008, 8.

91. Lorrie Grant, "Networks Hope Remote-Control Shopping Clicks," *USA Today*, May 25, 2005, 1B.

92. Laura Petrecca, "Interactive TV Ads Click with Viewers," *USA Today*, July 6, 2008, 1B.

93. Stephanie Clifford, "Cablevision Promises Interactive TV Ads—This Time for Sure," *New York Times*, September 15, 2009, https://mediadecoder.blogs.nytimes .com/2009/09/15/cablevision-promises-interactive-tv-ads-this-time-for-sure/.

94. Yuyu Chen, "Target Turns TV Viewership into a Shoppable Experience on Second-Screen," *Clickz*, March 24, 2014, https://www.clickz.com/target-turns-tv-viewership -into-a-shoppable-experience-on-second-screen/32550/; Todd Spangler, "NBCU Teams with Amex, Zeebox for T-Commerce," *Multichannel News*, November 9, 2012, https:// www.nexttv.com/news/nbcu-teams-amex-zeebox-t-commerce-306182.

95. George Winslow, "Tune in, Click through and Shop," *Broadcasting & Cable*, February 28, 2013, 28.

96. Anne Becker, "On Sale at Wisteria Lane," *Broadcasting & Cable*, September 18, 2006, 12.

97. Winslow, "Tune in, Click through and Shop."

98. Jason Lynch, "You Can Buy Everything You See on Lifetime & Wayfair's New TV Series," *Adweek*, October 19, 2016, https://www.adweek.com/convergent-tv/you-can -buy-everything-you-see-lifetime-wayfair-s-new-tv-series-174120/.

99. Ben Munson, "AT&T's Advertising Behemoth Is Coming for Facebook and Google," *FierceCable*, July 9, 2018, https://www.fiercecable.com/video/at-t-s-advertising -behemoth-coming-for-facebook-and-google.

100. Alex Kantrowitz, "Your TV and Phone May Soon Double Team You," *Advertising Age*, April 28, 2014, 8.

101. Laura Heller, "Amazon Makes Original Content Shoppable," *FierceRetail*, July 14, 2016, https://www.fierceretail.com/operations/amazon-makes-original-content-shoppable.

102. "Connecting Your Living Room for Shopping," CNBC, December 24, 2015, https://www.cnbc.com/video/2015/12/24/connecting-your-living-room-for-shopping.html.

103. George Winslow, "H&M Plans 'Shoppable' Super Bowl Ad," *Broadcasting & Cable*, January 6, 2014, https://www.nexttv.com/news/ces-hm-plans-shoppable-super-bowl-ad-128280.

104. Jeff Baumgartner, "Startup Eyes Deeper Connection to TV Advertising," *Broadcasting & Cable*, October 16, 2017, 40.

105. Terry Nguyen, "Online Shopping Changed, and We Barely Noticed," *Vox*, May 5, 2021.

106. Alice E. Marwick, "Instafame: Luxury Selfies in the Attention Economy," *Public Culture* 27, no. 1 (2015): 137–160; Brooke Erin Duffy and Emily Hund, "'Having It All' on Social Media: Entrepreneurial Femininity and Self-Branding among Fashion Bloggers," *Social Media + Society* 1, no. 2 (2015): 1–11; Brooke Erin Duffy, *(Not) Getting Paid to Do What You Love: Gender, Social Media, and Aspirational Work* (New Haven, CT: Yale University Press, 2017); Sophie Bishop, "Anxiety, Panic and Self-Optimization: Inequalities and the YouTube Algorithm," *Convergence* 24, no. 1 (2018): 69–84.

107. Shoshana Wodinsky, "TikTok Wants to Be QVC for Teens," *Gizmodo*, February 8, 2021, https://gizmodo.com/tiktok-wants-to-be-qvc-for-teens-1846220742.

108. Joseph Frydl, "How HSN and 'Eat Pray Love' Turn Content into Commerce," *Advertising Age*, August 4, 2010, https://adage.com/article/madisonvine-news/branded-content-lessons-hsn-eat-pray-love/145235.

109. See Amazon's StyleSnap at https://affiliate-program.amazon.com/resource-center/amazon-influencers-introducing-stylesnap (accessed April 4, 2022).

CHAPTER 8

Epigraph: Foreword to *Minutes of the First Meeting of the Operations Research Discussion Group, October 5, 1959* (New York: Advertising Research Foundation, 1959).

1. Melvin Anshen, "Management Science in Marketing: Status and Prospects," *Management Science* 2, no. 3 (1956): 229, 231, 223.

2. Peter Miller, "Governing by Numbers: Why Calculative Practices Matter," *Social Research* 68, no. 2 (2001): 379–396.

3. Sigfried Giedion, *Mechanization Takes Command: A Contribution to Anonymous History* (Minneapolis: University of Minnesota Press, 2013), 24.

4. Michel Callon, "Introduction: The Embeddedness of Economic Markets in Economics," *Sociological Review* 46, no. S1 (1998): 27.

5. *Are There Consumer Types? An Attempt to Predict Buying Behavior from Demographic and Personality Traits* (New York: Advertising Research Foundation, 1964), 1.

6. Harry Roberts, "The Role of Research in Marketing Management," *Journal of Marketing* 22, no. 1 (1957): 26. On attribution, see Harrison Smith, "People-Based Marketing and the Cultural Economies of Attribution Metrics," *Journal of Cultural Economy* 12, no. 3 (2019): 201–214.

7. Bruce G. Carruthers and Wendy Nelson Espeland, "Accounting for Rationality: Double-Entry Bookkeeping and the Rhetoric of Economic Rationality," *American Journal of Sociology* 97, no. 1 (1991): 31–69.

8. Alfred Kuehn, "How Advertising Performance Depends on Other Marketing Factors," *Journal of Advertising Research* 2, no. 1 (1962): 3.

9. Frank M. Bass and Leonard J. Parsons, "Simultaneous-Equation Regression Analysis of Sales and Advertising," *Applied Economics* 1, no. 2 (1969): 103.

10. Leo Bogart, "Is It Time to Discard the Audience Concept?," *Journal of Marketing* 30, no. 1 (1966): 51.

11. Frank M. Bass, "Some Case Histories of Econometric Modelling in Marketing: What Really Happened?," in *Proceedings of the First ORSA/TIMS Special Interest Conference on Market Measurement and Analysis*, ed. David B. Montgomery and Dick R. Wittink (Stanford, CA: Stanford University, 1979), 4.

12. Lawrence Friedman, "Game-Theory Models in the Allocation of Advertising Expenditure," *Operations Research* 6, no. 5 (1958): 699.

13. "Seminars Explore Ad Goals, Ratings," *Back Stage*, October 8, 1965, 3; emphasis added.

14. Dan Horsky, "An Empirical Analysis of the Optimal Advertising Policy," *Management Science* 23, no. 10 (1977): 1037.

15. Daniel Pope, *The Making of Modern Advertising* (New York: Basic Books, 1983), 284.

16. "Impact Measurement Only True Yardstick for Radio, Declares Ted Hill, WTAG," *Billboard*, July 23, 1949, 11.

17. Michael J. Naples, "Electronic Media Research: An Update and a Look at the Future," *Journal of Advertising Research* 24, no. 4 (1984): 39, 43.

18. For example, pollster George Gallup got his start measuring the impact of print ads, eventually working for Young & Rubicam. Sarah E. Igo, *The Averaged American: Surveys, Citizens, and the Making of a Mass Public* (Cambridge, MA: Harvard University Press, 2007); Daniel J. Robinson, *The Measure of Democracy: Polling, Marketing, and Public Life, 1930–1945* (Toronto: University of Toronto Press, 1999).

19. Leo Bogart, "Media Models: A Reassessment," *Journal of Advertising* 4, no. 2 (1975): 28–29.

20. *Minutes of the Second Meeting of the Operations Research Discussion Group, February 16, 1960* (New York: Advertising Research Foundation, 1960), 3.

21. *Minutes of the Second Meeting of the Operations Research Discussion Group*, 4–5. Halbert could not have been speaking for all of DuPont, which invested in extensive efforts to measure how its institutional advertising succeeded in educating the public about the virtues of big business. See Anna McCarthy, *The Citizen Machine: Governing by Television in 1950s America* (New York: New Press, 2010), chap. 1.

22. "Operations Research Discussion Group, Wednesday Afternoon Session, November 21, 1962," in *A Report of the Eighth Meeting of the ARF Operations Research Discussion Group* (New York: Advertising Research Foundation, 1962), 31. Agency personnel, even those in optimum-seeking departments, were not so inclined to dismiss communication goals. For example, BBDO's media director underscored advertising's communication function in a draft of one casually sexist speech: Herbert D. Maneloveg, "36-24-36, but No Body," Magazine Advertising Sales Club of New York, Sheraton East Hotel, May 26, 1965, 6.

23. Richard E. Quandt, "Estimating the Effectiveness of Advertising: Some Pitfalls in Econometric Methods," *Journal of Marketing Research* 1, no. 2 (1964): 51.

24. Horace C. Levinson, "Experiences in Commercial Operations Research," *Journal of the Operations Research Society of America* 1, no. 4 (1953): 234.

25. Joel Dean, "Cyclical Policy on the Advertising Appropriation," *Journal of Marketing* 15, no. 3 (1951): 268.

26. Sidney Hollander Jr., "A Rationale for Advertising Expenditures," *Harvard Business Review* 27, no. 1 (1949): 81–82, 87.

27. Samuel P. Hayes Jr., "Behavioral Management Science," *Management Science* 1, no. 2 (1955): 179; Richard D. McKinzie, "Oral History Interview with Samuel P. Hayes," July 16, 1975, https://www.trumanlibrary.gov/library/oral-histories/hayessp.

28. Wendell R. Smith, "Product Differentiation and Market Segmentation as Alternative Marketing Strategies," *Journal of Marketing* 21, no. 1 (1956): 7. Smith worked for the Alderson & Sessions consultancy, which was among the leaders in marketing OR.

29. William J. Baumol and Charles H. Sevin, "Marketing Costs and Mathematical Programing," *Harvard Business Review* 35, no. 5 (1957): 52–60; Charles H. Sevin, "Measuring the Productivity of Marketing Expenditures," in *Marketing and the Computer*, ed. Wroe Alderson and Stanley J. Shapiro (Englewood Cliffs, NJ: Prentice-Hall, 1963), 164–177.

30. Charles H. Sevin, "Sales Cost Accounting and the Robinson-Patman Act," *Journal of Marketing* 2, no. 3 (1938): 216.

31. Sevin, 218.

32. Baumol and Sevin, "Marketing Costs," 57, 52. Baumol was listed as a member of the ARF's OR discussion group from 1962 to 1964.

33. John D. C. Little and Russell Ackoff, "How Techniques of Mathematical Analysis Have Been Used to Determine Advertising Budgets and Strategies," *Proceedings of the 4th Advertising Research Foundations Conference* (1958): 19–23.

34. John F. Magee, "The Effect of Promotional Effort on Sales," *Journal of the Operations Research Society of America* 1, no. 2 (1953): 64–74; M. L. Vidale and H. B. Wolfe, "An Operations-Research Study of Sales Response to Advertising," *Operations Research* 5, no. 3 (1957): 370–381.

35. G. A. Bradford, "What General Electric Is Doing to Evaluate the Effect on Sales of Its Industrial and Consumer Advertising," in *Minutes of the Second Meeting of the Operations Research Discussion Group*, 1–2, 15, 13.

36. Kenneth Longman, "Marketing Effort—How Productive Is It?," in *Minutes of the Third Meeting of the Operations Research Discussion Group, Friday, April 22, 1960* (New York: Advertising Research Foundation, 1960), 9.

37. Kenneth Longman, "Marketing Effort—How Productive Is It?," in *Operations Research in Advertising: A Summary of Papers Presented to ARF's Operations Research Discussion Group* (New York: Advertising Research Foundation, 1961), 28.

38. Longman, 28–31.

39. Robin Kaiser-Schatzlein, "How Money Talks," *Mother Jones* 46, no. 5 (2021): 65.

40. "A Profit Yardstick for Advertising," *Business Week*, November 22, 1958, 50.

41. "Profit Yardstick for Advertising," 50.

42. "Profit Yardstick for Advertising," 53.

43. "Test Tube Sales," *Broadcasting*, December 31, 1962, 5.

44. John F. Lawrence, "Reappraising Ads," *Wall Street Journal*, May 28, 1963, 1. Both General Motors and DuPont reportedly spent $750,000 on this work in 1964. William S. Hoofnagle, "Experimental Designs in Measuring the Effectiveness of Promotion," *Journal of Marketing Research* 2, no. 2 (1965): 154.

45. Malcolm A. McNiven, "Choosing the Most Profitable Level of Advertising: A Case Study," in *How Much to Spend for Advertising?* (New York: Association of National Advertisers, 1969), 90.

46. McNiven, 96.

47. Philip Kotler, "Operations Research in Marketing," *Harvard Business Review* 45, no. 1 (1967): 187.

48. Arthur A. Porter, memorandum, December 7, 1956, 1, JWT Information Center Records, New York Media Department: Reports and Press Cuttings, 1956–1973, box 13, Hartman Center.

49. Porter.

50. Wroe Alderson, "Marketing and the Computer: An Overview," in *Marketing and the Computer* (Englewood Cliffs, NJ: Prentice-Hall, 1963), 5.

51. John D. C. Little, "Adaptive Experimentation," in *A Report of the Twelfth Meeting of the ARF Operations Research Discussion Group, April 19, 1965* (New York: Advertising Research Foundation, 1965), 10.

52. Little, 22.

53. Donald C. Marschner, "Theory versus Practice in Allocating Advertising Money," *Journal of Business* 40, no. 3 (1967): 292.

54. *A Report of the Eleventh Meeting of the ARF Operations Research Discussion Group, November 10, 1964* (New York: Advertising Research Foundation, 1964), 55.

55. Martin Kenneth Starr, "Management Science and Marketing Science," *Management Science* 10, no. 3 (1964): 568–569.

56. Kenneth A. Longman, "A Recent History of Media," *Management Science* 15, no. 10 (1969): B566, B567.

57. Roberts, "Role of Research," 25.

58. Quoted in Dallas W. Smythe, *Dependency Road: Communications, Capitalism, Consciousness, and Canada* (Norwood, NJ: Ablex Publishing, 1981), 33.

59. Charles E. Garvin, "The Thompson Viewpoint on Direct Mail," August 25, 1959, 2, JWT Information Center Records, Documents Relating to Media and Media Buying, 1957–1986, box 13, Hartman Center.

60. "Accountability Emerges as Factor in Agency Compensation: Harper to 4A's," *Advertising Age*, October 22, 1962, 1.

61. "Harper Proposes Agencies Share in Profits," *Broadcasting*, October 22, 1962 32.

62. Bill Harvey, "A Brief Personal History of Media Optimization" (unpublished manuscript, shared by the author, February 2018), 5.

63. John S. Wright, "Marion Harper, Jr.," *Journal of Marketing* 31, no. 1 (January 1967): 69–70.

64. Armand Mattelart, *Advertising International: The Privatisation of Public Space*, trans. Michael Chanan (London: Routledge, 1991).

65. Michael Farmer, *Madison Avenue Manslaughter*, 2nd ed. (New York: LID Publishing, 2017), 90–91.

66. Russell L. Ackoff and James R. Emshoff, "Advertising Research at Anheuser-Busch, Inc. (1963–1968)," *Sloan Management Review* 16, no. 2 (1975): 1–15.

67. Ackoff and Emshoff, 13.

68. Dana Canedy, "Stanley D. Canter, 75, an Advisor to Corporations," *New York Times*, May 3, 1999.

69. "NAB Convention: Computers Work Praised," *Broadcasting*, April 8, 1963, 66.

70. "'Admen Must Improve Measurement Justification of Efforts,' Jones Says," *Advertising Age*, October 1, 1962, 1, 120.

71. "Computers' Future," *Broadcasting*, January 14, 1963, 44.

72. "Tune down Gain, Urges Harper," *Broadcasting*, November 19, 1962, 52–54.

73. "The Profit Squeeze" (display ad), *Wall Street Journal*, April 6, 1961, 27.

74. "Media Departments—1966," *Sponsor*, October 23, 1961, 27.

75. "Advertising Enters the Age of Computers," *Sponsor*, January 29, 1962, 25.

76. *Wall Street Journal*, October 8, 1962, 13.

77. "Mathematical Programming for Better Selection of Advertising Media," *Computers and Automation* 10, no. 12 (1961): 20.

78. Comment from Joseph Kelly in "Discussion of Learner and Godfrey Papers," in *Mathematical Methods of Media Selection: A Report of the Sixth Meeting of the ARF Operations Research Discussion Group* (New York: Advertising Research Foundation, 1961), 4.

79. Comment from Benjamin Lipstein in "Discussion of Learner and Godfrey Papers," 4.

80. Lipstein comment, 4.

81. Milton Godfrey, "Media Selection by Mathematical Programming: Linear and Non-Linear Models," in *Mathematical Methods of Media Selection*, 2.

82. John D. C. Little and Leonard M. Lodish, "MEDIAC: An On-Line Media Selection System" (working paper 298-67, Alfred P. Sloan School of Management, Massachusetts Institute of Technology, November 1967), 1.

83. "An Integrated System for Media Planning," March 14, 1968, 8, JWT Information Center Records, Documents Relating to Media and Media Buying, 1957–1986, box 13, Hartman Center.

84. Thomas P. Hughes, *Networks of Power* (Baltimore: Johns Hopkins University Press, 1983), 22.

85. Philip Shabecoff, "Research in Revolution," *Sponsor*, January 28, 1963, 61.

86. Joseph St. George, "How Practical Is the Media Model?," *Journal of Marketing* 27, no. 3 (1963): 32–33.

87. Shabecoff, "Research in Revolution," 61.

88. "Tune down Gain," 54.

89. Little and Lodish, "MEDIAC," 4; John David, *Operational Research and Marketing* (London: J. Walter Thompson Company, 1967), 10, JWT Publications, box IL2, Hartman Center.

90. John D. C. Little, "Models and Managers: The Concept of a Decision Calculus," *Management Science* 16, no. 8 (1970): B483.

91. Michael L. Ray and Alan G. Sawyer, "Behavioral Measurement for Marketing Models: Estimating the Effects of Advertising Repetition for Media Planning," *Management Science* 18, no. 4 (1971): P73–P89; David A. Aaker, "Management Science in Marketing: The State of the Art," *Interfaces* 3, no. 4 (1973): 17–31.

92. John A. Howard, "Buyer Behavior and Related Technological Advances," *Journal of Marketing* 34, no. 1 (1970): 20.

93. Ray and Sawyer, "Behavioral Measurement," P74.

94. Michael L. Ray and Alan G. Sawyer, "Repetition in Media Models: A Laboratory Technique," *Journal of Marketing* 8, no. 1 (1971): 22.

95. "WSM Nashville," *Broadcasting-Telecasting*, December 10, 1945, 13.

96. "NBC Spot Sales," *Broadcasting-Telecasting*, January 28, 1946, 37.

97. "TvB Presentation, Member Additions Portend Adulthood," *Billboard*, June 4, 1955, 13.

98. Oscar H. Gandy Jr. and Charles E. Simmons, "Technology, Privacy and the Democratic Process," *Critical Studies in Mass Communication* 3, no. 2 (1986): 155–168; Erik Larson, *The Naked Consumer: How Our Private Lives Become Public Commodities* (New York: Penguin Books, 1994), chap. 8.

99. Peter M. Guadagni and John D. C. Little, "A Logit Model of Brand Choice Calibrated on Scanner Data" (working paper 1304-82, Alfred P. Sloan School of Management, Massachusetts Institute of Technology, May 1982), 3.

100. Edward Wallerstein, "Measuring Commercials on CATV," *Journal of Advertising Research* 7, no. 2 (1967): 19.

101. Naples, "Electronic Media Research," 45.

102. Little and Ackoff, "How Techniques of Mathematical Analysis Have Been Used," 22.

103. Malcolm A. McNiven, introduction to *How Much to Spend for Advertising?* 8.

104. John D. C. Little, "Interview by Robert Klein," September 4, 2014, https://www .informs.org/content/download/348559/3217056/file/Robert%20Klein%20Inter-views%20John%20D.%20C.%20Little,%20September%204,%202014.pdf.

105. John D. C. Little, "Decision Support Systems for Marketing Managers," *Journal of Marketing* 43, no. 3 (1979): 15.

106. Guadagni and Little, "Logit Model of Brand Choice," 2.

107. Little, "Decision Support Systems," 25.

108. Guadagni and Little, "Logit Model of Brand Choice," 1.

109. Guadagni and Little, 35.

110. Jack Honomichl, "Dawn of the Computer Age," *Advertising Age*, August 20, 1987, 117.

111. Dun & Bradstreet also owned the A. C. Nielsen Company, which was IRI's main competition in the market for retail data services. The deal was abandoned due to antitrust concerns. For more on BehaviorScan, see Larson, *Naked Consumer*, 138–145.

112. Gandy and Simmons, "Technology, Privacy and the Democratic Process."

113. Larson, *Naked Consumer*, 137.

114. Joe Coogle, "Data-Base Marketing," *Marketing & Media Decisions* 25, no. 1 (1990): 75.

115. Quoted in Gary L. Lilien and Philip Kotler, *Marketing Decision Making: A Model-Building Approach* (New York: Harper & Row, 1983), 524.

116. Patricia F. Phalen, "The Market Information System and Personalized Exchange: Business Practices in the Market for Television Audiences," *Journal of Media Economics* 11, no. 4 (1998): 17–34.

117. "Optimizers Redefine the Art of the Deal," *Advertising Age*, May 11, 1998.

118. Marc Gunther, "New Software Reprograms TV Advertising," *Fortune*, May 11, 1998, 27.

119. Erwin Ephron, "Ad World Was Ripe for Its Conversion to Optimizers," *Advertising Age*, February 22, 1999.

120. John Little and colleagues viewed clickstream data with the same enthusiasm as retail scanning data—as resources for modeling and predicting consumer choices. Randolph E. Bucklin et al., "Choice and the Internet: From Clickstream to Research Stream," *Marketing Letters* 13, no. 3 (2002): 245–258.

121. Philip M. Napoli, *Social Media and the Public Interest: Media Regulation in the Disinformation Age* (New York: Columbia University Press, 2019), 39.

122. Lorenzo Franceschi-Bicchierai, "Facebook Doesn't Know What It Does with Your Data, or Where It Goes: Leaked Document," *Motherboard*, April 26, 2022, https://www.vice.com/en/article/akvmke/facebook-doesnt-know-what-it-does-with -your-data-or-where-it-goes.

123. Charles K. Ramond, "How Advertising Became Respectable," *Journal of Marketing* 28, no. 4 (1964): 1–4. On how business forecasting both detects and directs trends, see Devon Powers, *On Trend: The Business of Forecasting the Future* (Urbana: University of Illinois Press, 2019).

124. "DataVision Killed the Demographic Star," TV of Tomorrow Show, New York City, December 7, 2017.

CONCLUSION

Epigraphs: Carolyn Marvin, "Fables for the Information Age: The Fisherman's Wishes," *Illinois Issues* 17 (1982): 19; Dallas W. Smythe, "New Directions for Critical Communications Research," *Media, Culture & Society* 6, no. 3 (1984): 205–217.

1. Jackson Lears, *Fables of Abundance: A Cultural History of Advertising in America* (New York: Basic Books, 1994), 251.

2. William Leiss, Stephen Kline, Sut Jhally, Jacqueline Botterill, and Kyle Asquith, *Social Communication in Advertising*, 4th ed. (New York: Routledge, 2018); Roland Marchand, *Advertising the American Dream: Making Way for Modernity, 1920–1940* (Berkeley: University of California Press, 1985); Michael Schudson, *Advertising, the Uneasy Persuasion: Its Dubious Impact on American Society* (New York: Basic Books, 1984); Russell W. Belk and Richard W. Pollay, "Images of Ourselves: The Good Life in Twentieth Century Advertising," *Journal of Consumer Research* 11, no. 4 (1985): 887–897.

3. James V. Cook, "1970—Can Marketing Measure Up?," in *Effective Marketing Coordination: Proceedings of the Forty-Fourth National Conference of the American Marketing Association*, ed. George L. Baker, Jr. (Chicago: AMA, 1961), 21–22, 25–27. For a much different vision of economic management that was forced into "freedom" by a US-backed military coup, see Eden Medina, *Cybernetic Revolutionaries: Technology and Politics in Allende's Chile* (Cambridge, MA: MIT Press, 2011).

4. Mark Andrejevic, *iSpy: Surveillance and Power in the Interactive Era* (Lawrence: University Press of Kansas, 2007).

5. Paul Edwards, *The Closed World: Computers and the Politics of Discourse in Cold War America* (Cambridge, MA: MIT Press, 1996).

6. See, e.g., Mark Andrejevic, "Public Service Media Utilities: Rethinking Search Engines and Social Networking as Public Goods," *Media International Australia* 146, no. 1 (2013): 123–132; Antoine Haywood and Victor Pickard, "Public Access Television Channels Are an Untapped Resource for Building Local Journalism," *Nieman Lab*, November 10, 2021, https://www.niemanlab.org/2021/11/public-access-television-channels-are-an-untapped-resource-for-building-local-journalism/.

7. Dan Schiller, "Restructuring Public Utility Networks: A Program for Action," *International Journal of Communication* 14 (2020): 4994.

8. Sanjay Jolly and Ellen P. Goodman, *A "Full Stack" Approach to Public Media in the United States* (Washington, DC: German Marshall Fund of the United States, 2021).

9. Schiller, "Restructuring Public Utility Networks," 4995.

10. Robert W. McChesney, *Digital Disconnect: How Capitalism Is Turning the Internet against Democracy* (New York: New Press, 2013); Astra Taylor, *The People's Platform: Taking Back Power and Culture in the Digital Age* (New York: Metropolitan Books, 2014); Jessa Lingel, *The Gentrification of the Internet: How to Reclaim Our Digital Freedom* (Berkeley: University of California Press, 2021).

11. Christopher R. Martin, *No Longer Newsworthy: How the Mainstream Media Abandoned the Working Class* (Ithaca, NY: Cornell University Press, 2019).

12. C. Edwin Baker, *Advertising and a Democratic Press* (Princeton, NJ: Princeton University Press, 1994), 66–68.

13. Todd Gitlin, *The Whole World Is Watching: Mass Media and the Making and Unmaking of the New Left* (Berkeley: University of California Press, 1980); Victor Pickard, *Democracy without Journalism? Confronting the Misinformation Society* (New York: Cambridge University Press, 2019); Rachel Kuo and Alice Marwick, "Critical Disinformation Studies: History, Power, and Politics," *Harvard Kennedy School Misinformation Review* 2, no. 4 (2021): 1–12.

14. Marvin, "Fables for the Information Age," 24.

INDEX

Page numbers followed by *f* and *t* indicate figures and tables, respectively.